Programming PICs in BASIC

8-Pin Projects – Volume 1

Published by Electronic Products.

The publisher offers special discounts on bulk orders of this book.

For information contact:

Electronic Products
P.O. Box 251
Milford, MI 48381
www.elproducts.com
chuck@elproducts.com

Printed in the United States of America
Cover design by Rich Scherlitz

Table of Contents

Introduction

I've been programming Microchip PIC microcontrollers for years. When I started programming, the most common software option was assembly language programming which can be very confusing to the beginner. Then one day I discovered a product that advertised programming Microchip PIC® microcontrollers in BASIC. It was the PICBASIC compiler from microEngineering Labs. They eventually released a more powerful version called PICBASIC PRO and I have been using that compiler ever since. Along the way they developed a sample version of the PICBASIC PRO compiler that is limited to a few parts and only 31 commands but it is more than enough for someone to try out this great method of programming for free.

This book is a beginner's guide to using the PICBASIC PRO compiler with the 8-pin PIC12F683 microcontroller. To save cost for the reader I use the sample version of PICBASIC PRO for all the projects which you can download for free from www.melabs.com. It also is included with the low cost programming kit from www.CHIPAXE.com that I for the projects. This way the reader/student can complete the projects and learn right along with me for little investment. This language is so powerful, the 31 commands will not seem like a limitation until you get comfortable with this method of programming and want to do more complex projects. This book is not intended to be a complete source of knowledge on programming with PICBASIC PRO or the PIC12F683 but by the end of the chapters and projects you should be able to branch out and develop your own more complex projects.

It may lead to a profession in programming and that may force you to learn other languages such as the C Language I use in my book "Beginner's Guide to Embedded C Programming". Programming in PICBASIC PRO is very powerful though and can handle just about any task you need. Programming in PICBASIC PRO is also so easy, users from all skill levels can make it work well. After all

BASIC stands for Beginner All-purpose Symbolic Instruction Code and the key word is Beginner. BASIC is also great way to introduce microcontroller programming to many skill levels. You cannot read an electronics hobbyist magazine without seeing a project that is microcontroller based. The evolution of the homemade robot has advanced due to the advancements and affordability of microcontrollers. But for some it's still difficult to get started or to find all the right pieces you need to build your own microcontroller development lab. This book is intended to help you get started and guide you along the way.

If you have any questions regarding this book or the projects in this book, you can usually get me via email at chuck@elproducts.com. My website at www.elproducts.com shows all the books I've written to help readers get started programming. Now lets get started learning how to program Microchip PICs in BASIC.

Chapter 1 – Microcontroller Fundamentals

Many of the programming books I've read tend to skip over a couple of important topics; how a microcontroller works and how to use them. Many of the books go into great detail on the software or the project operation but assume you know all the steps to get that program into the microcontroller or exactly what a microcontroller is. I don't assume the reader knows any of this so in this first chapter I cover the fundamentals of Microcontrollers.

What is a Microcontroller?

Everybody reading this has probably used a Personal Computer (PC) run by a microprocessor which is often times an Intel microprocessor from the Intel Corporation. The PC's central microprocessor has several support items that allow it to function: 1) The memory, where programs are stored, known as a hard drive or ROM, 2) The RAM, or temporary memory used by the programs running in the microprocessor and 3) The interface to the outside world, also known as the BIOS or input and output (I/O) control. The PC's system mother board will connect these three components and also have a power supply and a system clock typically running at gigahertz speed and advertised as such to imply how fast the PC is (i.e. 1 Ghz processor).

The PC sends information through the I/O to be displayed on the monitor or printed paper via a connected printer. The I/O also reads the keyboard and mouse position. Basically everything the PC does with a useful purpose to humans runs through the I/O. All these components I've described make up the central processing unit or CPU of your home computer and is typically packaged in a large metal box most people just call "the computer". Now lets assume you could shrink all those components; microprocessor, ROM, RAM, I/O and system clock into a single integrated circuit.

It can be done and has been done. It's called a microcontroller and runs many of the products you already are familiar with such as mp3 players, cell phones or video game players.

A microcontroller is a miniature computer in a single integrated circuit with a small amount of ROM and RAM and lots of I/O. Figure 1-1 shows some of the various sized microcontrollers from Microchip Technology Inc. They offer a full line of microcontrollers that they call the PIC® microcontrollers. They offer many more than shown here but these are very common packages that range from an 8 pin small package up to a 40 pin large package with lots of I/O.

Figure 1-1: Microchip PIC Microcontrollers

How Does a Microcontroller Work?

A microcontroller requires a series of coded electrical charges stored into its ROM to control the micro's I/O. These electrical charges take on two states; high voltage or low voltage. You can think of these charges just like a light switch where it's either on or off. These charges can also be represented mathematically by the binary number system which only includes the digits 1 and 0. These individual charges are often called bits. The microcontroller is designed to do different operations on the I/O depending on the arrangement of these bits. An eight bit microcontroller uses 8 different charges combined to determine the operation to perform. This is also known as a byte value. Figure 1-2 shows an 8-bit byte value. There are 256 different possible arrangements of bits in an 8-bit value. Each one can represent a different operation in the microcontroller.

1	0	0	1	0	0	1	0

Figure 1-2: 8-bit byte value

When you have multiple bytes put together it is known as software or program code. When a microcontroller is said to be programmed or have code burned into it, it is getting these coded electrical signals stored into its ROM.

To function or run the code, the microcontroller needs a way to select each command from ROM one at a time, which is referred to as running a program. To do this, the microcontroller requires a clock oscillator, often times this is created with a separate part called a crystal or resonator connected to the microcontroller to create a continuous internal pulse train that drives the microcontroller's central circuitry. This is very similar to the PC Gigahertz speed clock. Gigahertz is one billion pulses per second. The speed in a microcontroller is much slower in the megahertz or million pulses per second. For most applications this is plenty fast enough.

When the micro is first powered up, the oscillator clock starts pulsing the same way our heart pulses our blood through our body. On each pulse of the clock, the micro retrieves a new command from ROM to execute on the I/O. By arranging these binary codes properly you can make the I/O pins switch on and off to control other electrical circuitry connected to the I/O pins. That circuitry could be a simple relay that turns a light on during the night and off during the day or it could be more complex and control the motors of a robot while reading an obstacle sensor. All you need to do is write the series of binary codes properly, which is the software. To make it easier to develop this binary code, compilers were developed. The PICBASIC PRO language compiler is a BASIC language compiler.

What is a Basic Language Compiler?

You could create software by individually setting or clearing each bit in memory but most microcontrollers (PIC's included) offer a software creation tool called assembly language. Assembly language uses short little acronyms to represent various simple operations in the microcontroller and the assembler built into this software will convert those acronyms into the 1's and 0's. Assembly language can be considered a very cryptic language to most people and because of this compilers were created. A compiler is a PC software application that converts easy to read and understand words (BASIC commands) into an assembly language file and then lets the assembler covert the result into the binary code (1's and 0's) the micro needs. The PICBASIC PRO compiler from microEngineering Labs is a BASIC language compiler for the Microchip 8-bit microcontrollers. Binary code is the lowest level of software and BASIC is considered a high level language. Once that binary code file is created, then the microcontroller can be programmed. The binary file will have the suffix .hex (i.e. program.hex).

The PICBASIC PRO compiler is very powerful and the full version sells for around $250. You can use the full version for developing complex projects and products. In this book though, we will use the sample version of the PICBASIC PRO compiler which is limited to selected PIC microcontrollers and 31 command lines. This may not sound like a lot of capability but you will soon be surprised by how much we can accomplish in 31 commands. The best part is the sample version is free to download so you can try out their great compiler by recreating the projects in this book.

How does the Microcontroller actually get programmed?

A microcontroller programming tool is a custom designed module that receives the binary code file created by the compiler and generates the electrical signals the microcontroller needs to see. The binary file is downloaded to the program memory or ROM of the device through specific I/O pins on the microcontroller. There are many different types of programming tools and are typically just called a Microchip PIC Programmer.

Figure 1-3 shows the PICkit™2 Starter Kit which is a complete starter kit for programming Microchip Technology microcontrollers. The development board included with the PICkit 2 Starter Kit makes it easy to get started programming Microchip PIC microcontrollers. The development board has a 20 pin socket and comes with a PIC16F690 microcontroller. The development board can be powered from the PICkit™2 programmer or separately. The PICkit™2 gets its power from the PC USB port so the development board power is limited to about 50 milliamps. If you needed more power then you can power the board separately. The development board has four LEDs, a momentary switch and a potentiometer wired directly to the 20 pin socket. This can offer a quick way for a beginner to get started.

Figure 1-3: PICkit™2 Starter Kit

Microchip shares the schematics and software for the PICkit 2 programmer so there are also many clone versions of the PICkit 2 programmer that you can purchase from various sources. The clone version shown in Figure 1-4 is a shrunken version of the PICkit 2. It has the USB connector built into the end and a ribbon cable with the programming connection wires at the other. This eliminates the need for a separate USB cable and is very compact. I will use this clone in most of the experiments in this book and any future project books that follow since it's part of a nice little development system called the CHIPAXE programming system (chipaxe.com).

Figure 1-4: CHIPAXE PICkit 2 Clone Programmer

This programmer can actually plug into the same development board as the PICkit 2 Starter Kit or any In-Circuit Serial Programming board but before I get into that let me explain In-Circuit Serial Programming.

In-Circuit Serial Programming

The PICkit 2 programmer has only six connections to the development board for powering the board and for programming the PIC microcontroller. Two of the connections are power and ground. Some programmers like the CHIPAXE clone PICkit 2 only have 5 connections. For those that have six, the extra pin is for calibrating the internal oscillator on some PIC's. This is rarely needed so many clones leave it off. This leaves three connections that all programmers have and are specifically for programming the code into the microcontroller. All of these programmers use these three pins in a serial method of programming called In-Circuit Serial Programming (ICSP). Having only a few connections to your circuit board can be very handy so I wanted to explain how to use the ICSP feature.

The main advantage to ICSP is the ability to program the PIC in-circuit. It does this through three main connections which are also

I/O pins that you may want to use in your project. The biggest hang-up you may have with ICSP is the serial communication signal can get affected by the circuitry connected to the PIC I/O. For example, the three connections used to program a PIC in-circuit are: Vpp (MCLR pin), Data (PGD pin) and Clock (PGC pin). If the Clock or Data signals are not able to send the correct signal, the PIC will not program properly and you will get a verify error. If you build your own development board or connect circuitry to the programming pins of the PIC then you may affect these signals. There are recommended connection methods from Microchip to get around this interference so let me explain those.

ICSP Hardware Interface

The schematic in Figure 1-5 shows the ICSP connections and all the possible connection issues to watch out for in your design. Because of the way the ICSP feature works, you don't want to add any capacitance to the programming connections since this can delay the signals. Even the capacitance on the Vdd line should be monitored per the PIC programming specification. The PIC programmer actually cycles the Vdd line off and on while sending the Vpp signal to the MCLR pin. This is done to put the PIC in programming mode. If there is too much capacitance, it may slow the signal down and not meet the programming specs. You can get the programming specs for any PIC microcontroller at the Microchip.com website.

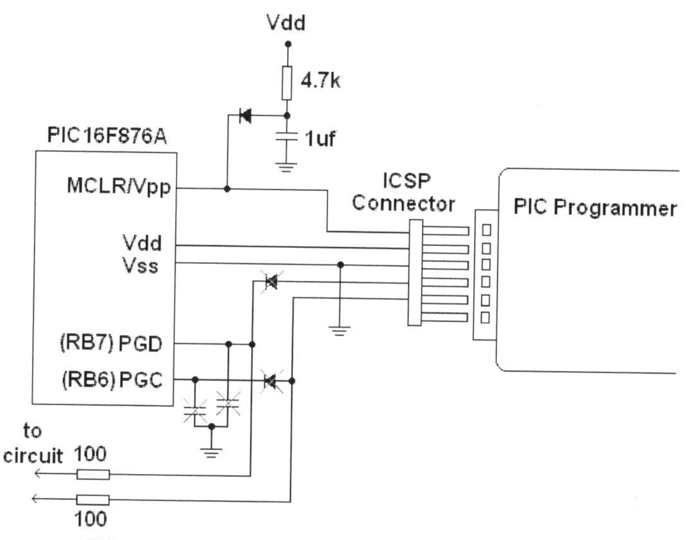

Figure 1-5: PICkit 2 Hardware Interface

You also don't want to load down the clock or data signal. The components that are crossed out show what could affect the signal. The diodes on the Data and Clock lines are a mistake because two-way communication occurs when programming and verifying the part. These are pretty easy to see why they should not be included in your design.

What isn't quite so clear is the diode between the MCLR reset circuit and the MCLR/Vpp pin. This is recommended because the PIC programmer sends a high voltage, low current signal to the Vpp line of around 12v -13.5v for a short period of time. You don't want that signal feeding into your Vdd regulator so the diode helps protect for that. This is actually just an extra safety precaution though because the current entering the MCLR pin is extremely small and the MCLR pull-up resistor will knock it down to prevent any damage. In most cases you can get by without the diode.

Another recommendation which is often missed is the series resistors on the PGD and PGC lines between the PIC and the rest of the circuit. These isolate your circuit from the PGD and PGC signals so your circuit doesn't load down the PIC programmer. This is usually where a problem may occur with ICSP. One hundred ohm resistors should not affect your circuit function but it should be plenty of resistance to isolate the programmer.

Another option to get around all this is to add a switch to your circuit. A four pole switch that allows you to disconnect the PGC, PGD, Vdd and MCLR pins from the circuit during programming prevents the rest of the connected circuitry from loading down clock and data and disconnects the MCLR resistor from the programming connection. Figure 1-6 shows the schematic for that type of arrangement. This is a premium type of setup and most development boards won't offer this switched connection arrangement.

What is interesting is the development board included with the PICkit 2 Starter Kit only includes some of these protections. You have to make sure you don't add circuitry to the data and clock lines that might affect the programming operation. Therefore it is best to add the 100 ohm resistors to your circuit if you use those I/O pins in your design.

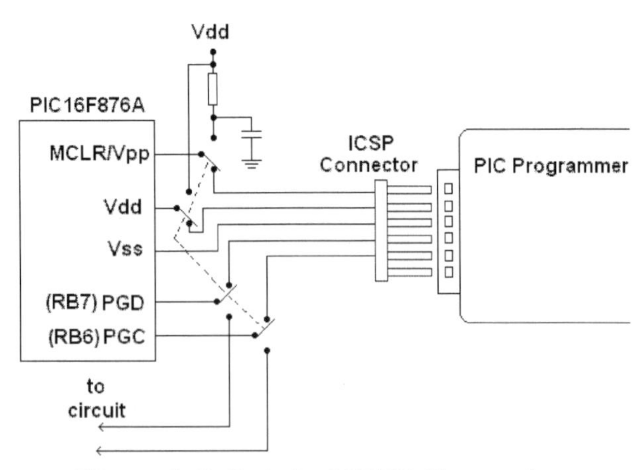

Figure 1-6: Switched ICSP Connection

ICSP Pin-Out

To be helpful I will give you some of the ICSP pins for the various PIC microcontroller DIP packages. Table 1 below summarizes it for you. This should save you the trouble of looking through all those data sheets.

ICSP Connection	40 pin pin #	28 pin pin#	20 pin pin#	18 pin pin#	14 pin pin#	8 pin pin#
Vpp\MCLR	1	1	4	4	4	4
Vdd	11 & 32	20	1	14	1	1
Vss	12 & 31	8 & 19	20	5	14	8
Data	40	28	19	13	13	7
Clock	39	27	18	12	12	6

Table 1: ICSP Connection Pin Numbers

Figure 1-7 shows a simple 40 pin development board I made to test out ICSP with the PICkit 2 programmer. It's not pretty but it worked. I even added a couple LEDs so I could test it with a few flash LED programs. The 16F887 part used here has an internal oscillator so I didn't need to add an external resonator. I prefer to use a breadboard setup for most of my projects so I can easily move wires around. I often just use a breadboard module like the one shown in Figure 1-8.

Figure 1-7: Simple Development Board uses ICSP

The small board shown plugged into the breadboard is called a CHIPAXE module. CHIPAXE modules are built with the ICSP connectors on the end to make it real easy to plug a PIC microcontroller into a breadboard and still program with the PICkit 2 or the CHIPAXE PICkit 2 clone programmer. I'll use this CHIPAXE programming system for the projects in this book. The boards don't have a 100 ohm resistor on the PGC or PGD pins so you may have to add those when you use those pins in your circuit. The projects in this book will use some resistance when necessary to eliminate the ICSP interference concern.

The CHIPAXE programmer is a 5 volt output only programmer that draws power from the USB port. The programmer can supply up to 50 milliamps to your circuit. You can therefore power your circuit directly from the programmer so you don't need an external power supply. If you exceed the 50 milliamps then your project will start to act strange and your PC USB port may shut down. In that case you need to supply power externally. The CHIPAXE programmer should sense that and stop supplying power.

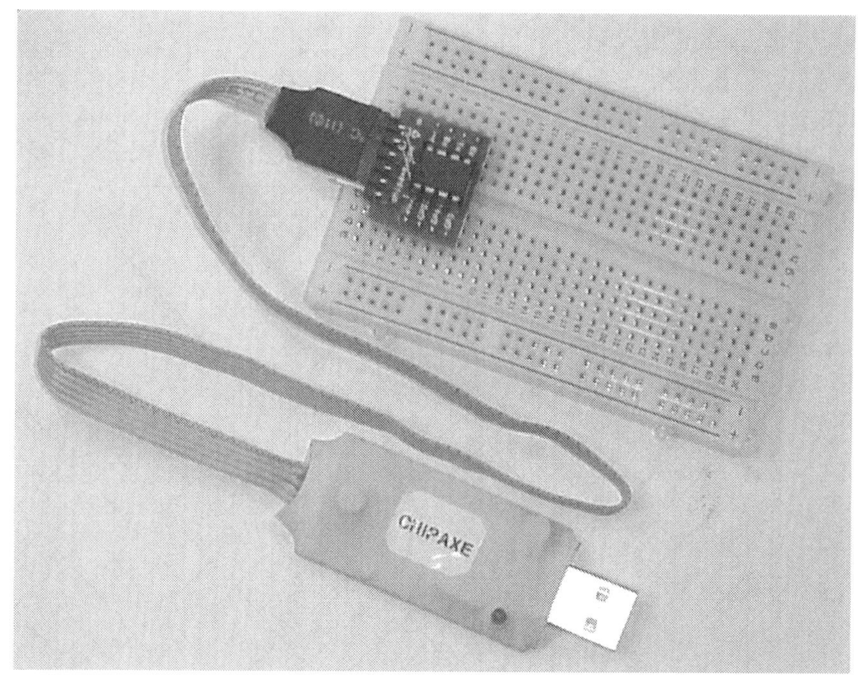

Figure 1-8: CHIPAXE Breadboard Programming System

Installing the Software

Now you are ready to get your own programming setup installed on your computer so you can complete the projects in this book. The PICBASIC PRO compiler sample version comes with its own windows editor for writing the programs and sending the binary file to the programmer. The editor is called the MicroCode Studio. The PICkit 2 programmer software has a command line option which allows us to use the PICkit 2 or CHIPAXE within the MicroCode Studio editor to one click compile and program the PIC microcontroller using the ICSP. We have to set all this up on your PC before we can create the first project. The steps we will take are:

1) Download all the files needed from my website at www.elproducts.com/eightpinbook_vol1.htm.
2) Install the PICBASIC PRO compiler sample version.
3) Install MicroCode Studio Software (this is part of the PICBASIC PRO installation).
4) Setup the PICkit 2 interface and command line software within Microcode Studio.

Installing MicroCode Studio and PICBASIC PRO

Download all the files for this installation on to your computer. You will need a PC running windows XP or Vista for best performance. You can get the PICBASIC PRO sample version at:

www.melabs.com/pbpdemo.htm

The full set of project files can be downloaded at my website:

www.elproducts.com/eightpinbook_vol1.htm.

You need to run the PICBASIC PRO installation file so run the PBPDEMO4.exe file to install the sample version of the PICBASIC PRO compiler. The compiler will step you through several screens shown in Figures 1-9A through 1-9G.

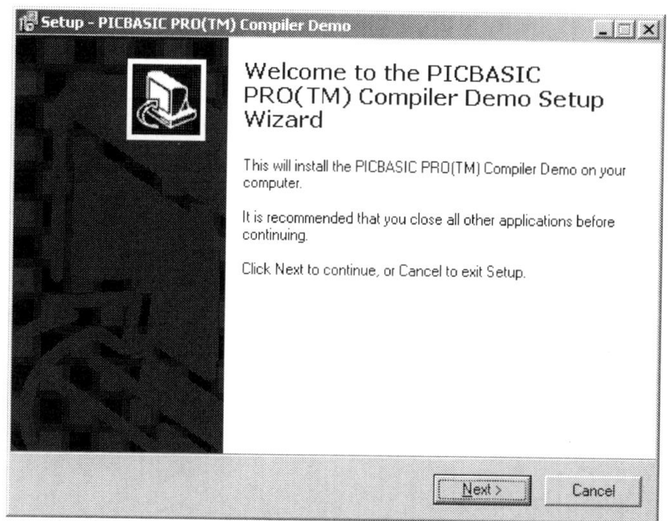

Figure 1-9A: First PICBASIC PRO Installation Screen

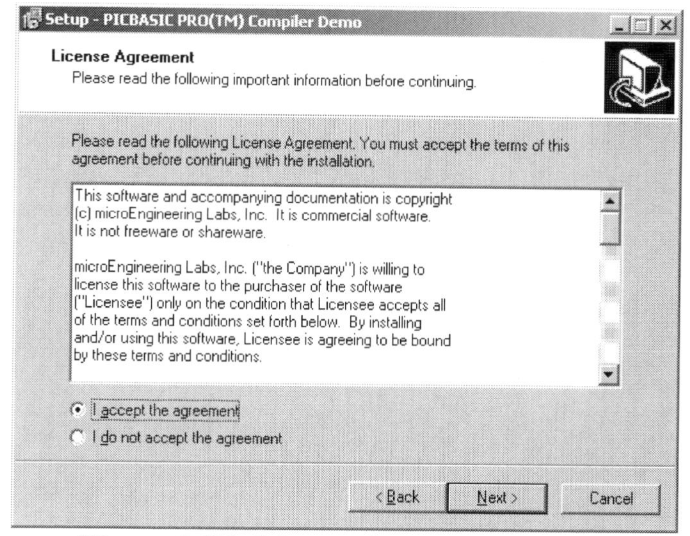

Figure 1-9B: License Agreement Screen

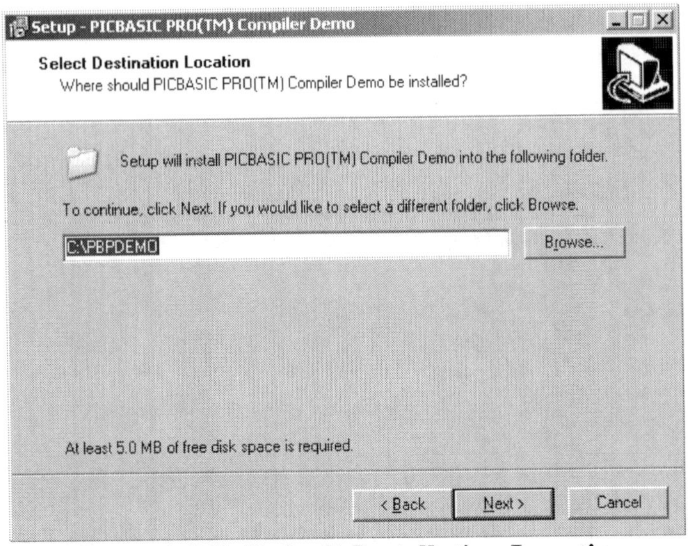

Setup - PICBASIC PRO(TM) Compiler Demo
Select Destination Location
Where should PICBASIC PRO(TM) Compiler Demo be installed?
Setup will install PICBASIC PRO(TM) Compiler Demo into the following folder.
To continue, click Next. If you would like to select a different folder, click Browse.
C:\PBPDEMO
Browse...
At least 5.0 MB of free disk space is required.
< Back
Next >
Cancel

Figure 1-9C: Default Installation Location

Setup - PICBASIC PRO(TM) Compiler Demo
Select Start Menu Folder
Where should Setup place the program's shortcuts?
Setup will create the program's shortcuts in the following Start Menu folder.
To continue, click Next. If you would like to select a different folder, click Browse.
PICBASIC PRO Compiler Demo
Browse...
Don't create a Start Menu folder
< Back
Next >
Cancel

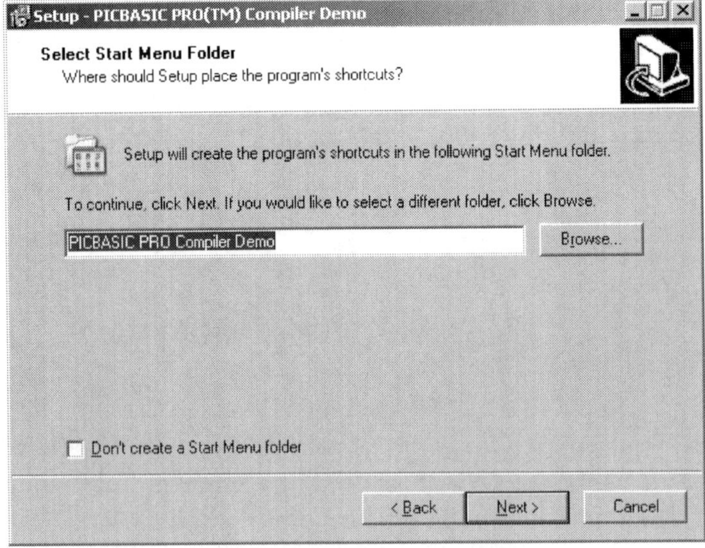

Figure 1-9D: Windows Start Menu Title

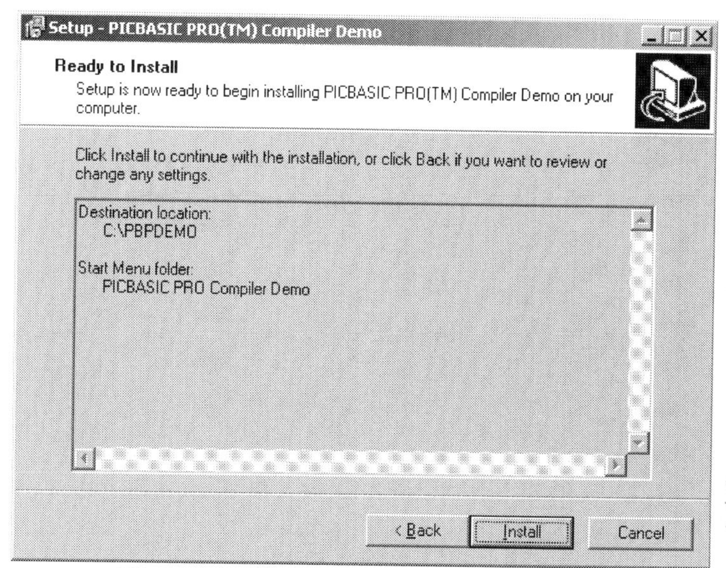

Figure 1-9E: Ready to Install Screen

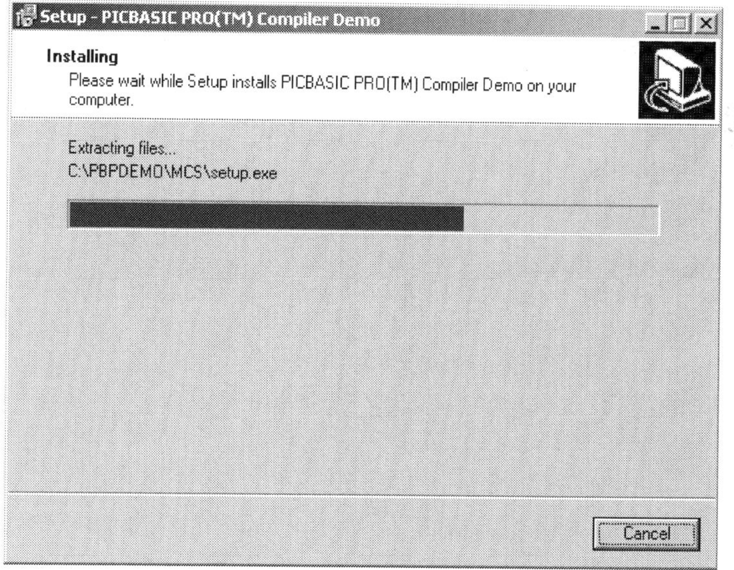

Figure 1-9F: Installation in Progress

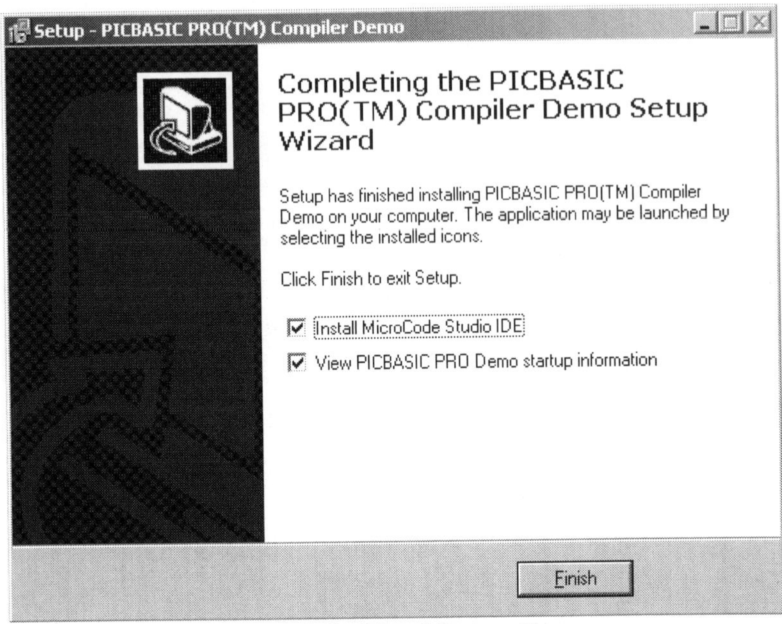

Figure 1-9G: Final Setup Screen

When the PICBASIC PRO installation is complete the last screen (Figure 1-9G) will offer to "Install MicroCode Studio IDE". Make sure that option is checked before clicking on the "Finish" button. The MicroCode Studio will install automatically after you press Finish.

The MicroCode Studio will begin and step you through the screens in Figure 1-10A through 1-10F

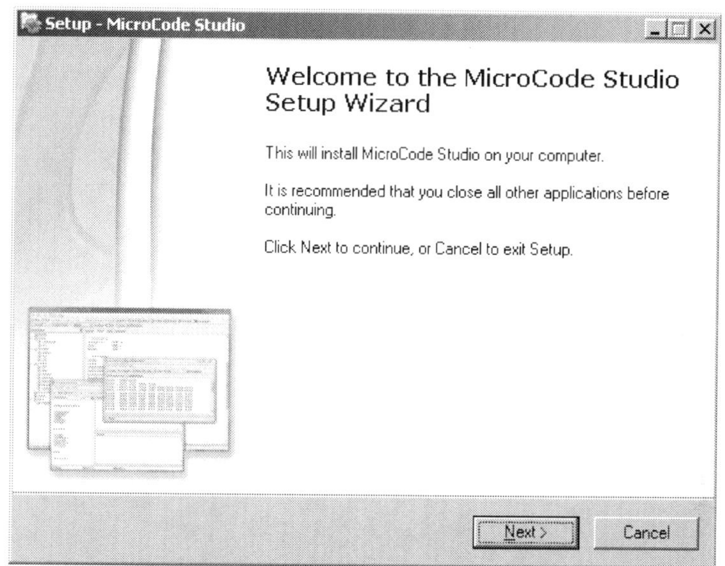

Figure 1-10A: First MicroCode Studio Screen

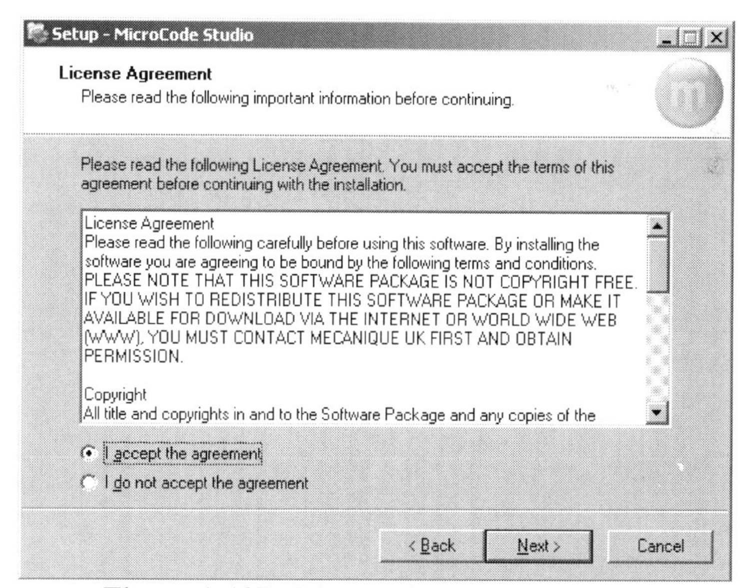

Figure 1-10B: License Agreement Screen

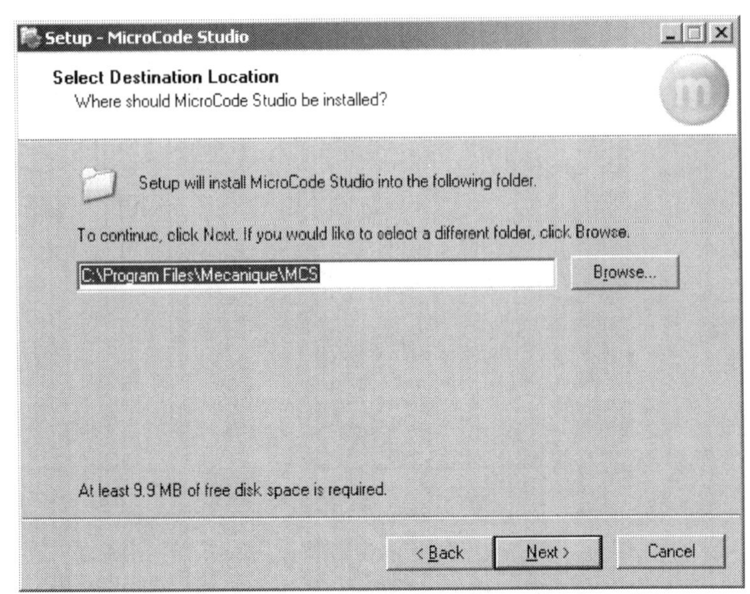

Figure 1-10C: Default Location Installation Screen

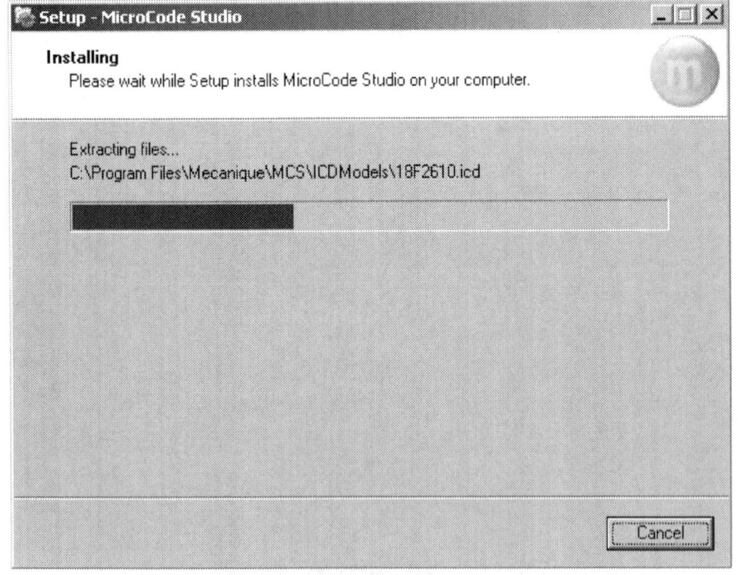

Figure 1-10D: Installation in Progress Screen

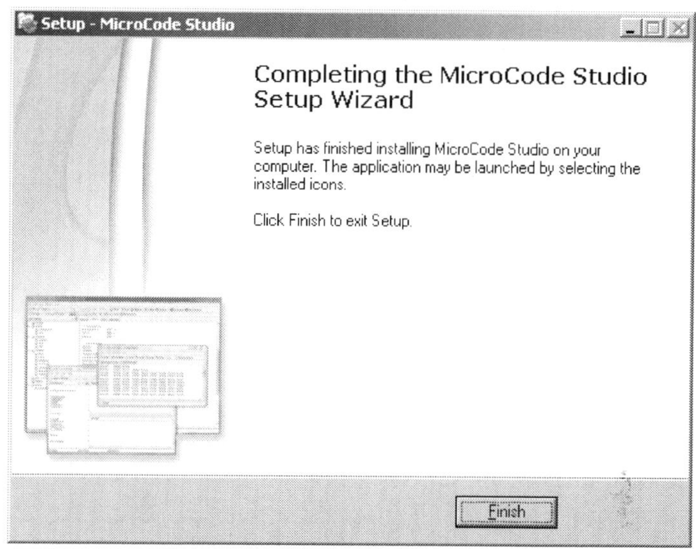

Figure 1-10E: Final MicroCode Studio Screen

After installation is complete, start Microcode studio and it should look similar to the picture in Figure 1-10F below.

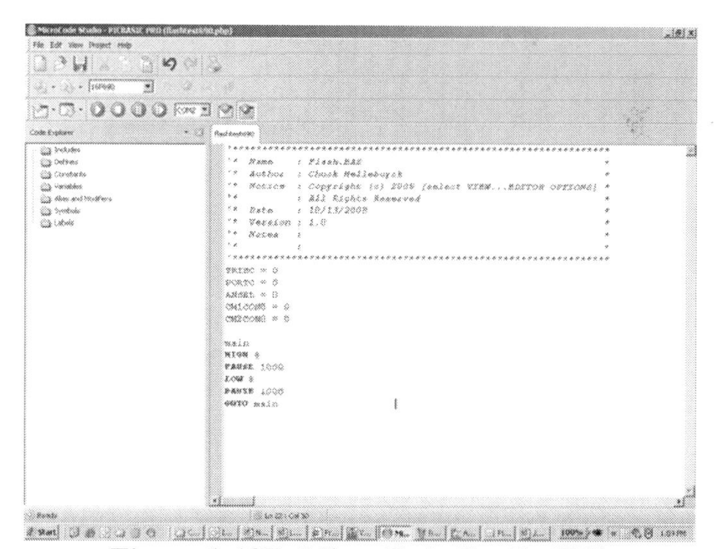

Figure 1-10F: MicroCode Studio IDE

Programmer Setup

Using the PICkit 2 or CHIPAXE PICkit 2 clone programmer within MicroCode Studio requires a few key steps including the command line parameter setup. First make sure you have the PICkit 2 command line executable "pk2cmd.exe" file installed on your computer. It was part of the package of files you can download from my website at the link below. www.elproducts.com/eightpinbook_vol1.htm.
You can also download it at www.microchip.com/pickit2. The latest version, as I write this, is V1.20.

When you have that installed on your computer open the MicroCode Studio screen and then click on the "View > Compile and Program Options" selection as shown in Figure 1-11 or click on the little arrow next to the controller selection window as shown in Figure 1-12 . The window in Figure 1-13 should appear.

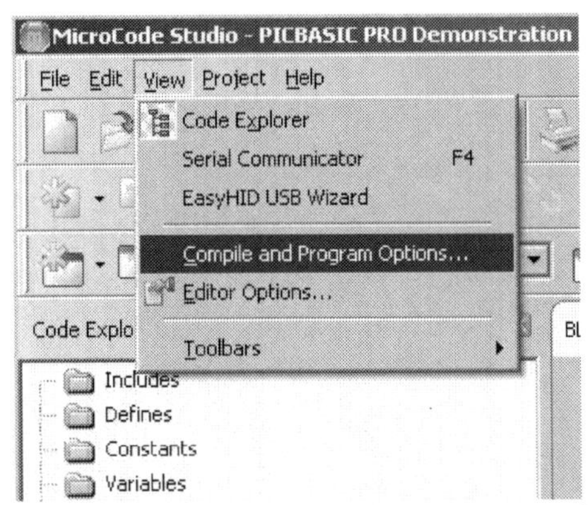

Figure 1-11: Programmer Setup Menu Option

The default programmer shown is microEngineering Labs own programmer which is another PIC programmer you can consider though for the projects in this book we'll be using the CHIPAXE PICkit 2 clone programmer which works with the pk2cmd.exe command line interface. We need to create a new programmer entry to use the CHIPAXE.

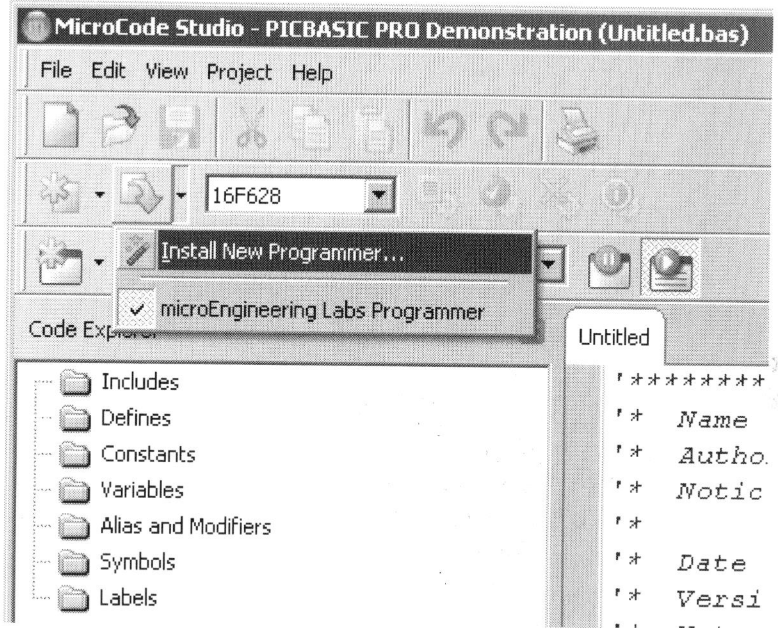

Figure 1-12: Programmer Setup Window

Click on the “Install New Programmer” button to create the new setup in MicroCode Studio. The “Create a custom programmer entry” should already be selected so click on the “Next >” button.

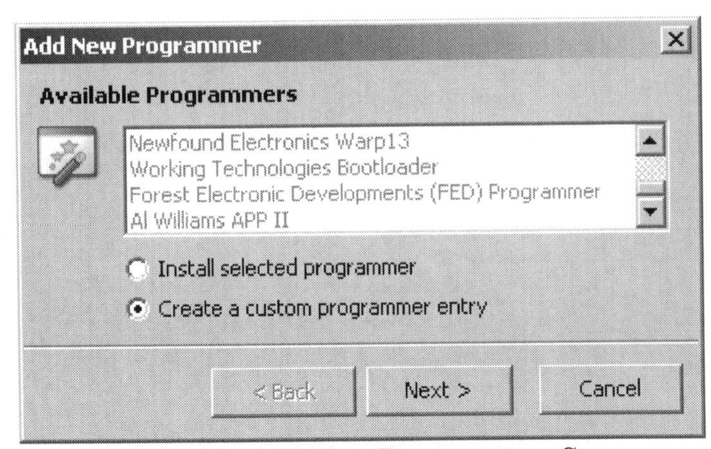

Figure 1-13: Add New Programmer Screen

A second "Add New Programmer" window will appear with a blank line. Enter the name you want to appear in the programmer selection window when you select your programmer at compile time. I chose to call it simply "CHIPAXE" as you can see in 1-14. This same setup will work with either an actual PICkit 2 or the CHIPAXE programmer.

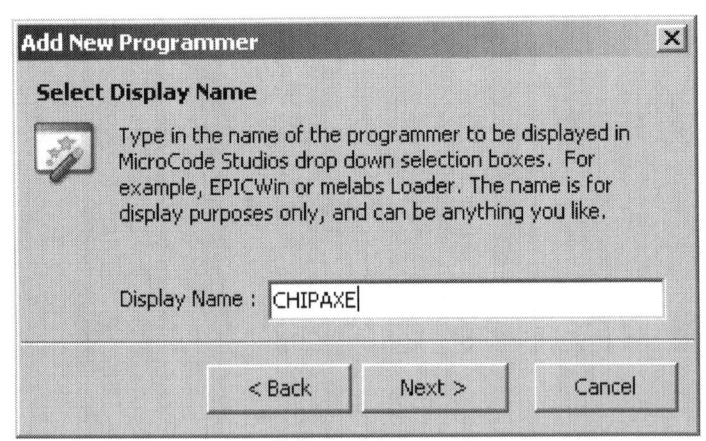

Figure 1-14: Enter PICkit2 in Window

After you enter the programmer name, click the “Next >” button and the “Add New Programmer” window will ask for the programmer command line executable. The command line executable for the CHIPAXE is the pk2cmd.exe file. Enter this into the “Programmer Filename:” window as seen in Figure 1-15.

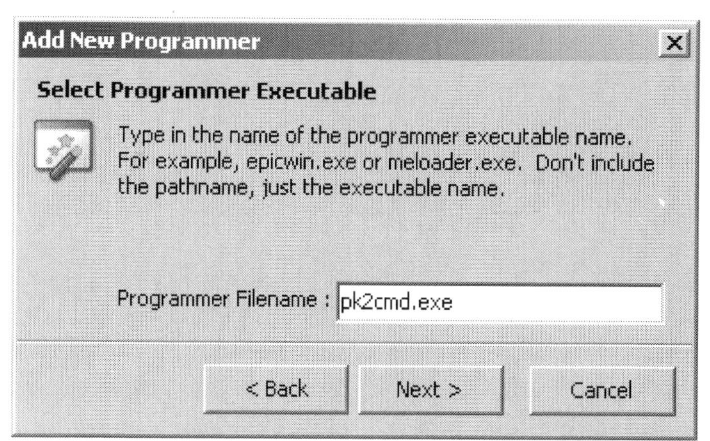

Figure 1-15: Enter PICkit2 Command Line

You don’t have to know where the file is for the next step but when you click on the “Next >” button, the screen in Figure 1-16 will pop up asking for the file location. You have a choice in how you want to find the “pk2cmd.exe” executable file. You can have MicroCode Studio search for the command line automatically or you can manually select it yourself if you remember where you put it on your computer. The manual option is a lot faster.

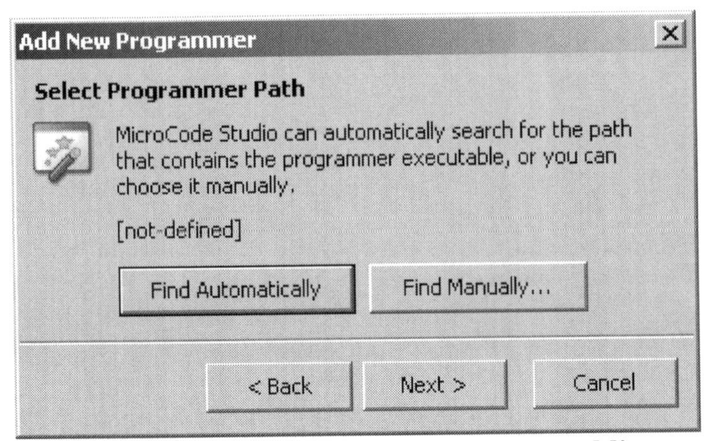

Figure 1-16: Find the PICkit2 Command line

The next step I'm going to describe is the most critical and the most difficult because these are the command line options that will transfer the PIC microcontroller part you are using and the binary .hex file the compiler creates to the PICkit2 prior to starting the PICkit2 command line executable. The command line options include both PICkit2 format and MicroCode Studio format options combined. Enter the command line exactly as you see it here with spacing and capitalization. If you are an advanced PC user then you may modify the command line from what I show below depending on how you want to use the PICkit2. The "Readme for PK2CMD.txt" file that downloaded with the "pk2cmd.exe" file explains all the PICkit2 command line options. I highly suggest you read that over before changing from what I show below.

/PPIC$target-device$ /F$long-hex-filename$ /M /R /T /H3

Figure 1-17 shows the command line entered in the MicroCode Studio screen. Let me explain my choices for those that want to understand how it works.

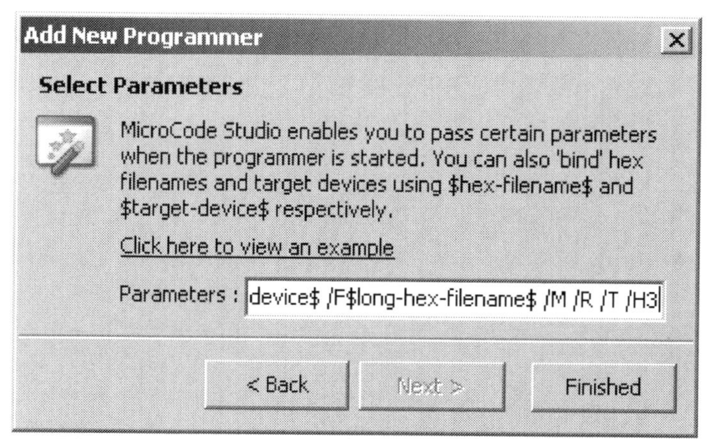

Figure 1-17: Command Line Option Window

The **/P** option selects the PIC but I wanted MicroCode Studio to pass that on from its own selection window. MicroCode Studio offers that detail as a parameter "$target-device$" but it doesn't include the PIC in front of the number (i.e. 16F690 instead of PIC16F690) so you have to type that in. This gives us the **"/PPIC$target-device$"** section of the command line option.

I'm using the PICBASIC PRO compiler and the **/F** option selects the .hex file the compiler created. MicroCode Studio offers the path to the compiled and assembled .hex file through a parameter called "$long-hex-filename$". This part of the command line thus becomes **"/F$long-hex-filename$"**.
The rest of the command line options are where you may modify the setup that I have. I first added the **/M** option which directs the CHIPAXE to erase and then program all the memory locations including program memory, the configuration, the EEPROM and the ID memory.

The **/R** directs the CHIPAXE programmer to release the MLCR line or reset line after programming. This prevents the programmer from holding the development board in reset mode.

The **/T** option tells the CHIPAXE programmer to power the development board from the USB port. Remember you only have 50 milliamps to use with this.

/H3 is the final option I chose. When the programmer command line starts operating, a pop up window will appear and show the programming operation running. I like to keep that window open for a few seconds after the programming is complete. The /H3 keeps it open for three seconds. Figure 1-18 shows a typical pop up window.

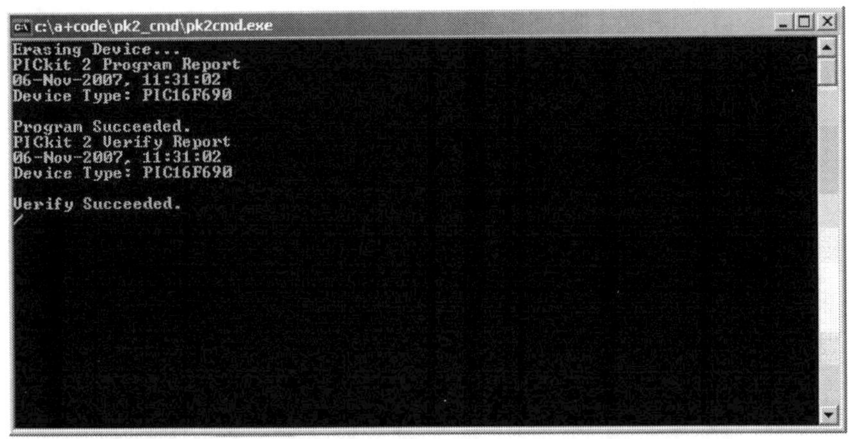

Figure 1-18: pk2cmd.exe Programming Screen

Special Note:
If you try to program your first part and the programming screen pops up quickly and then disappears, then you probably have a mistake in your command line or the pk2cmd.exe cannot find the PICkit 2 programmer. This could be a problem with the USB port so make sure you have everything connected properly.

After you setup the CHIPAXE programmer you also need to verify PICBASIC PRO can be found by the MicroCode Studio. Click on the “View > Compile and Program Options” in the menu bar and you should see the screen in Figure 1-19. The path to where PICBASIC PRO is stored on your computer should show up as seen in Figure 1-19. Verify this is correct for your setup otherwise MicroCode Studio will give you an error. This should be automatically set up correctly by the installation but it’s best to check.

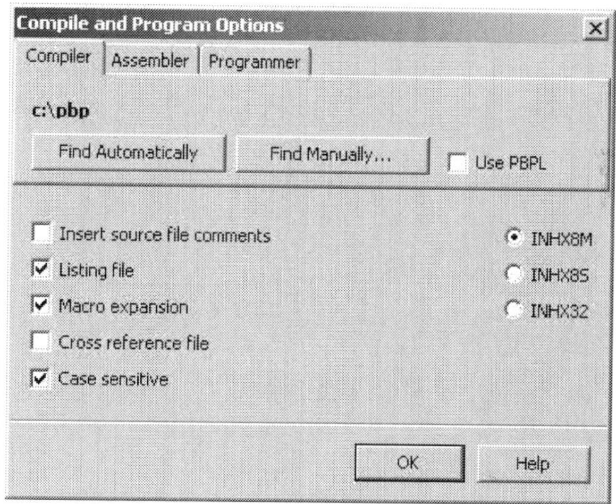

Figure 1-19: PICBASIC PRO Setup

That’s it. Follow these steps and you should have your CHIPAXE programming system up and running within MicroCode Studio with the PICBASIC PRO compiler and have a one click, compile and program solution.

Configuration Settings

There is one final step you need to understand before we can create our first program. Outside the structure of your program, the PIC has certain bits that are set at program time to control the special internal settings of the microcontroller. These settings control the internal watchdog timer, power-up timer, and oscillator selection and a few more.

All of the options for the part can be seen in the PIC data sheet under the CONFIG register section. The PICBASIC PRO compiler puts those configuration settings in a separate file that it calls at compile time. The settings file will be in a .inc file that has the name of the PIC you are using. The projects in this book will use the PIC12F683 eight pin microcontroller. In this case, the configuration file is named 12F683.inc. You will find it in the PBPDEMO directory, where you installed the PICBASIC PRO compiler. The file will contain two separate configuration lines.

device pic12F683, intrc_osc_noclkout, wdt_on, mclr_on, protect_off

and

__config _INTRC_OSC_NOCLKOUT & _WDT_ON & _MCLRE_ON & _CP_OFF

This line in the file is where the PICBASIC PRO compiler gets the information on how to set the configuration bits inside the .hex file. In this example, the internal RC oscillator is used as the system clock. This is the setting we will use in the book projects. You can open the 12F683.inc file with notepad and modify it if necessary.

If you needed to use the MCLR pin as a digital I/O pin then you would change the **mclr_on** to **mclr_off** and **MCLRE_ON** to **MCLRE_OFF**. You can then save the file for any future builds.

For the projects in this book you shouldn't need to change the default settings but if you want to change them in the future you now know where to look.

Chapter 2 – Flash an LED

Now we can write our first program that will simply flash an LED connected to the GP0 pin of the 12F683 eight pin microcontroller. This is a simple project but proves out the whole process of writing software, programming the microcontroller and watching the application run.

Figure 2-1 shows the completed project built into a breadboard. The CHIPAXE programmer powers the board and the red LED will flash at a rate of ½ second on and ½ second off.

Figure 2-1: Final Flash LED Project

Hardware

The hardware is built on a breadboard that has letters lined up with the column of connections and numbers for the rows. The connections can be reproduced based on the table of connections below. Figure 2-2 also shows the schematic for this project.

Connection Table

```
Micro         - Pin 1 at C6
Vdd Jumper    - a6 to +rail
Vss Jumper    - j6 to -rail
Green Jumper- j7 to j12
330 ohm       - i12 to i18
Red LED       - Anode j18, Cathode -rail
```

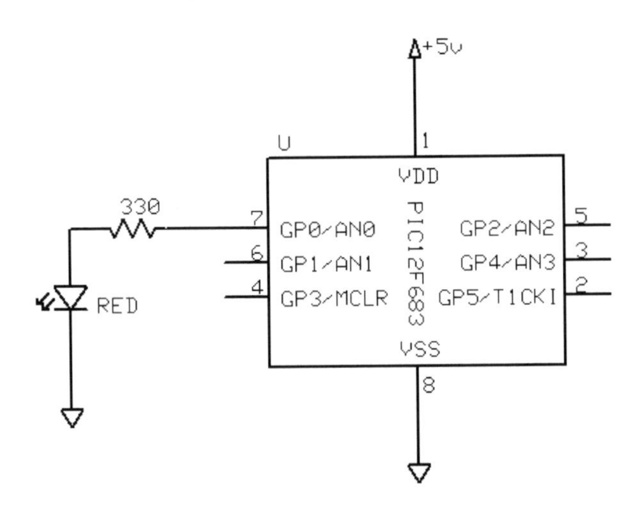

Figure 2-2: Flash LED Schematic

Software

```
'****************************************************
'*  Name    : Blink.BAS
'*  Author  : Chuck Hellebuyck
'*  Notice  : Copyright (c) 2009 Electronic Products
'*          : All Rights Reserved
'*  Date    : 8/20/2009
'*  Version : 1.0
'*  Notes   :
'*          : CHIPAXE-8 Pin 1 at C6
'*          : Vdd Jumper - a6 to +rail
'*          : Vss Jumper - j6 to -rail
'*          : Green Jumper - j7 to j12
'*          : 330 ohm - i12 to i18
'*          : Red LED - Anode j18, Cathode -rail
'****************************************************

 ANSEL = 0   ' Set I/O to digital
 CMCON0 = 7 ' Comparator off

 main:
 HIGH GPIO.0    'LED on
 PAUSE 500      'Delay 1/2 second
 LOW GPIO.0     'LED off
 PAUSE 500      'Delay 1/2 second
 GOTO main      'Loop back and do it again
```

The software is quite simple because it uses some of the commands that make PICBASIC PRO easier to use that many other compilers. The first section is required to setup the I/O as digital. When the PIC12F683 is first powered up, the I/O defaults to analog mode. The I/O pins share a connection to both analog and digital features. The ANSEL = 0 sets all the I/O pins to digital mode.

```
ANSEL = 0  ' Set I/O to digital
```

The 12F683 also has an internal comparator which can be shut down with the CMCON0 = 7 line.

```
CMCON0 = 7 ' Comparator off
```

The main program loop begins with the label "main" followed by a semi-colon. We will use this marker in a future GOTO command.

```
main:
```

The I/O pins on the 12F683 are referred to as General Purpose Input Output pins or GPIO. The internal register that controls these pins individually is also called the GPIO register. These can be controlled by writing to the GPIO register directly but we would also need to setup the TRISIO register inside the 12F683. The TRISIO determines if the pins are a digital input pin or a digital output pin. Both of these are automatically controlled with the **HIGH** or **LOW** commands in PICBASIC PRO.

The GPIO.0 is the nickname for the GP0 pin. The software uses the HIGH command to place a high signal on that pin. This will light the LED. PICBASIC PRO doesn't care if you use capitals or small case letters. The MCStudio should recognize the command an automatically capitalize the command.

```
HIGH GPIO.0   'LED on
```

The next command is the **PAUSE** command. This command just creates a ½ second delay as the value 500 represents 500 milliseconds or ½ second.

```
PAUSE 500     'Delay 1/2 second
```

The program then turns the LED off by setting the same pin low.

```
LOW GPIO.0    'LED off
```

The same pause command line delays another ½ second.

```
PAUSE 500    'Delay 1/2 second
```

The program then uses the GOTO command to jump back to the main label and repeat the operation over again.

```
GOTO main    'Loop back and do it again
```

How to Program using CHIPAXE:

1) Enter the software listing into the MicroCode Studio editor window.
2) Make sure Target Processor window in MCStudio shows "12F683" (Figure 2-3).
3) Connect the CHIPAXE programmer to the development board with the PIC12F683 in its socket. The dotted side of the ribbon cable should match up to the arrow on the CHIPAXE board (Figure 2-4).
4) Click on the little arrow next to the compile/program button (it's next to the target processor window from step 2) and make sure "CHIPAXE" is selected (Figure 2-5).
5) Click on the compile/program button to program the CHIPAXE module (Figure 2-5).

This should compile your program and bring up the pk2cmd.exe command line pop-up window. You should see it program and then complete the process. The pop-up window should close after a three second delay. The red LED should be blinking on the development board. If you don't get this working, go back through the steps and see if you missed something. Getting a simple LED to flash is a great first project.

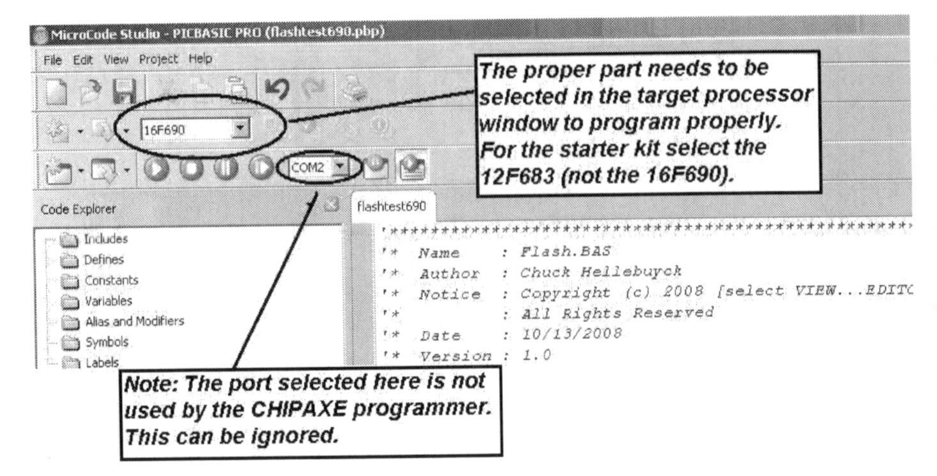

Figure 2-3: Target Processor Window

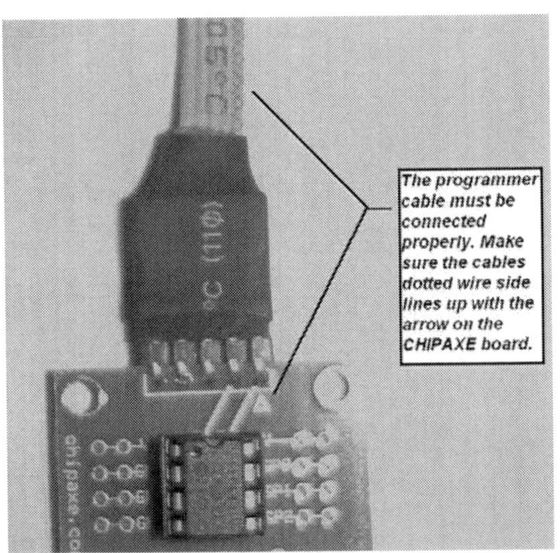

Figure 2-4: Connection to the CHIPAXE board

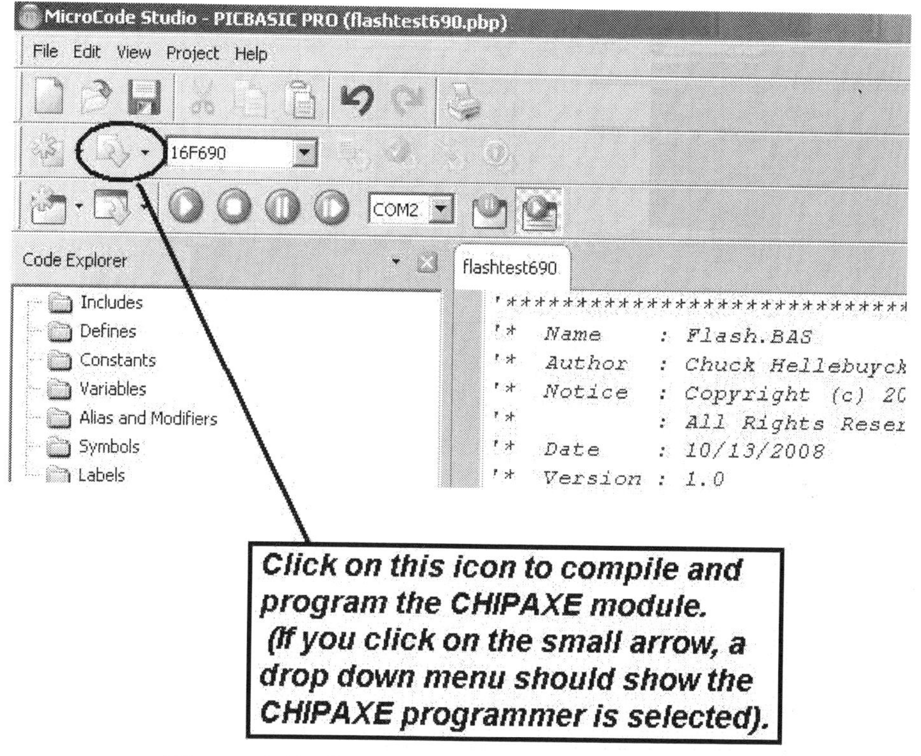

Figure 2-5: Compile and Program Icon

Next Steps

Simple next steps are to change the pause value to a lower number to flash the LED faster. You could also connect the LED to a different pin and then change the number in the high and low command lines to make that new connection pin flash the LED.

Questions

1) Which command line makes the I/O pins digital?

   A) CMCON0 = 7
   B) ANSEL = 0
   C) ANSEL = 7
   D) CMCON = 0

2) Which command line delays for 250 milliseconds

   A) HIGH 250
   B) PAUSE 500
   C) PAUSE 250
   D) LOW 250

3) The CMCON = 7 command line turns the comparator on.

   A) True
   B) False

4) GPIO.0 represents which 12F683 pin ?

   A) GP3
   B) GP1
   C) GP5
   D) GP0

Chapter 3 – LED Traffic Light

This chapter expands on the project in the previous chapter but in a more useful way. This project reproduces a traffic light with a Red, Yellow and Green LED. The timing of each LED is separately controlled which allows you to control how long each color is lit. The project also shows how to control multiple LEDs from one microcontroller. The completed project is shown in Figure 3-1.

Figure 3-1: Final LED Traffic Light Project

Hardware

The hardware builds off the previous project by just adding the two extra LEDs. The unique operation is in the software. The connection table for the breadboard describes the connections and the schematic is shown in Figure 3-2.

Connection Table

```
Micro         - Pin 1 at C6
Vdd Jumper    - a6 to +rail
Vss Jumper    - j6 to -rail
Green Jumper- j7 to j12
330 ohm       - i12 to i18
Red LED       - Anode j18, Cathode -rail
Green Jumper- i8 to i13
330 ohm       - i13 to i20
Yellow LED    - Anode j20, Cathode -rail
Blue Jumper  - f9 to f15
330 ohm       - g15 to g23
Green LED     - Anode j23, Cathode -rail
```

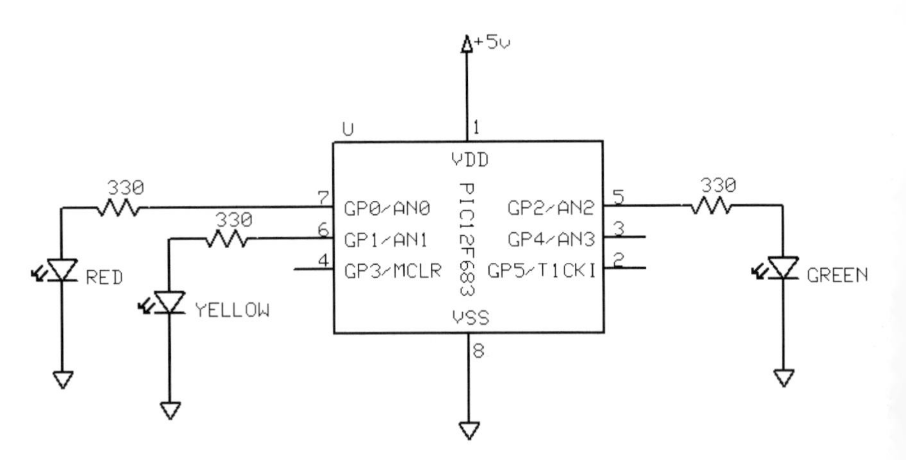

Figure 3-2: LED Traffic Light Schematic

Software

The software starts off by setting the ports to digital and the comparator shut down. Then the main label is added again to mark the beginning of the main loop of code.

```
ANSEL = 0  ' Set I/O to digital
CMCON0 = 7 ' Comparator off

main:
```

The Red LED is lit first for 2 seconds and then shut off to create the stop light portion of the traffic light. All actions are on the GP0 pin which is connected to the Red LED.

```
HIGH GPIO.0      'Light Red LED
PAUSE 2000       'Delay 2 second
LOW GPIO.0       'Red LED off
```

The Green LED is next and lit for the same amount of time as the Red LED. The Green LED is connected to the GP2 pin.

```
HIGH GPIO.2      'Green LED on
PAUSE 2000       'Delay 2 second
LOW GPIO.2       'Green LED off
```

Finally the Yellow LED is lit for one second and then off. The Yellow is connected to the GP1 pin.

```
HIGH GPIO.1      'Yellow LED on
PAUSE 1000       'Delay 1 second
LOW GPIO.1       'Yellow LED off
```

The final step is to loop back to the main label to light the Red LED and repeat the process.

```
GOTO main        'Loop Back to Red LED
```

Software Listing

```
'****************************************************
'*  Name    : Traffic.BAS
'*  Author  : Chuck Hellebuyck
'*  Notice  : Copyright (c) 2009 Electronic Products
'*          : All Rights Reserved
'*  Date    : 8/20/2009
'*  Version : 1.0
'*  Notes   :
'*          : CHIPAXE-8 Pin 1 at C6
'*          : Vdd Jumper - a6 to +rail
'*          : Vss Jumper - j6 to -rail
'*          : Green Jumper - j7 to j12
'*          : 330 ohm - i12 to i18
'*          : Red LED - Anode j18, Cathode -rail
'*          : Green Jumper - i8 to i13
'*          : 330 ohm - i13 to i20
'*          : Yellow LED - Anode j20, Cathode -rail
'*          : Blue Jumper - f9 to f15
'*          : 330 ohm - g15 to g23
'*          : Green LED -  Anode j23, Cathode -rail
'****************************************************

 ANSEL = 0  ' Set I/O to digital
 CMCON0 = 7 ' Comparator off

 main:
 HIGH GPIO.0    'Light Red LED
 PAUSE 2000     'Delay 2 second
 LOW GPIO.0     'Red LED off
 HIGH GPIO.2    'Green LED on
 PAUSE 2000     'Delay 2 second
 LOW GPIO.2     'Green LED off
 HIGH GPIO.1    'Yellow LED on
 PAUSE 1000     'Delay 1 second
 LOW GPIO.1     'Yellow LED off
 GOTO main      'Loop Back to Red LED
```

Next Steps

The logical next step is to change the delays to make the traffic light fit your application. If you wanted to make a real traffic light from a project like this then the delays need to be a lot longer. You can increase the pause delay to 65,535 or roughly 65 seconds. There are three unused I/O pins so adding three more LEDs for another crossing lane of traffic would be an option but unfortunately the GP3 pin is an input only pin and cannot drive an LED. This is an example of where a larger pin PIC would be the best choice.

Questions

1) Which is the correct way to create a label called main?

A) Main:
B) main:
C) MAIN:
D) All of the above

2) GP3 cannot drive an LED because?

A) It doesn't have enough current drive
B) It is an input only pin
C) It is connected to MCLR
D) It cannot be controlled by software

3) A PIC12F683 output pin can drive how much current?

A) 10 milliamps max
B) 20 milliamps max
C) 25 milliamps max
D) 5 milliamps max

4) To use the internal oscillator with GP4 as an I/O we set the configuration to?

A) intrc_osc_noclkout
B) intrc_oscillator
C) internal_oscillator
D) internal_osc

Chapter 4 – Sensing a Switch

On many projects you will need some kind of human interface to control the operation. A momentary push button switch is a very common way to do that. It can start and stop the operation or it could speed up or slow down what the microcontroller is controlling. In order to do that though the software needs to recognize that a switch was pressed. This project shows a simple method of sensing a momentary push button switch.

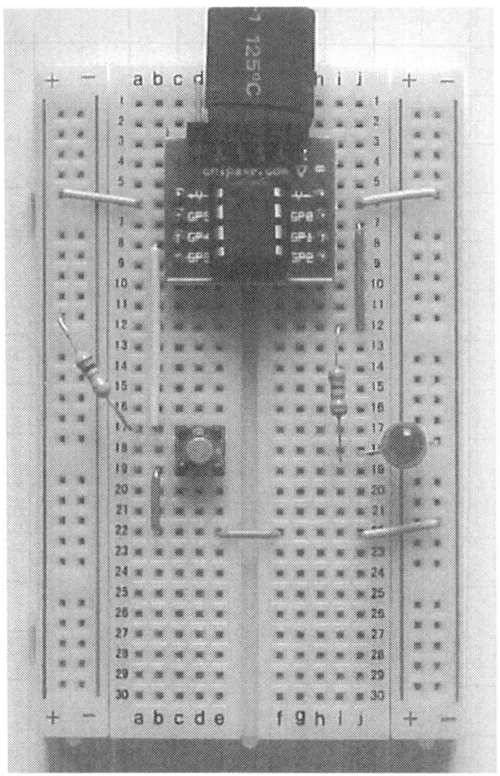

Figure 4-1: Final Switch Sensing Project

The software will have to monitor the switch continuously as part of the main loop of code and then respond. In this project the software will light an LED until the switch is pressed at which point the LED will shut off. As long as the switch is pressed the LED will stay off. As soon as the switch is released the LED will once again light up. The completed project is shown in Figure 4-1.

Hardware

The hardware uses the same Red LED connections as the first project we completed. The addition of the switch is shown in the schematic of Figure 4-2. The switch is wired as a low side switch meaning the circuit has a pull-up resistor to 5 volts so the input the micro is high when the switch is not pressed and low when the switch is pressed. This is known as a low side switch. If the parts were reversed and the switch was connected to 5 volts and the resistor to ground then it would be a high side switch.

The software will test the input pin GP4 to see if it changes to low indicating the switch has been pressed. The software will change the GP4 pin to an input. The connection table for the breadboard is below.

Connection Table

```
Micro             - Pin 1 at C6
Yellow Jumper     - a6 to +rail
Yellow Jumper     - j6 to -rail
Green Jumper      - j7 to j12
330 ohm           - i12 to i18
Red LED           - Anode j18, Cathode -rail
Yellow Jumper     - j22 to -rail
Orange Jumper     - f22 to e22
Orange Jumper     - b22 to b19
White Jumper      - b8 to b17
10k ohm           - a17 to +rail
Switch            - d17 to d19
```

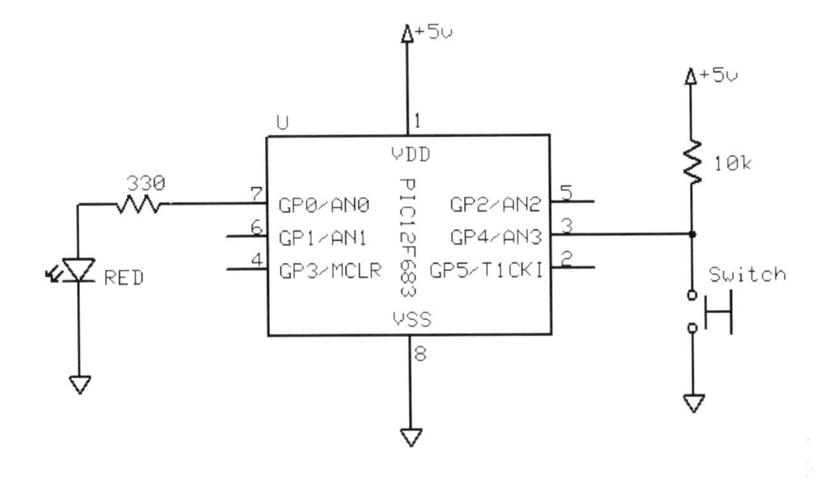

Figure 4-2: Switch Sensing Schematic

Software

The software starts off with the same I/O setup but now adds a new line to make the GP4 pin an input for the switch. This command acts on the internal TRISIO register of the PIC12F683. That register determines if the I/O pin is an input or an output. Each bit of that 8-bit register represents a pin. A 1 in the GP4 slot makes it an input and a 0 makes it an output. The INPUT GPIO 4 command line sets the GP4 pin to a 1 for input mode.

```
ANSEL = 0   ' Set I/O to digital
CMCON0 = 7 ' Comparator off
INPUT GPIO.4
```

The main label starts us off again which is followed by an IF-THEN-ELSE command from PICBASIC PRO. This command will test the equation after the IF to determine if it is true or false. If the equation is true then the command line following the THEN command line will get executed. If it's not true then the command following the ELSE line will get executed. In this case if the GP4 pin is high meaning the switch is not pressed then the Red LED is lit. If instead the GP4 pin is low meaning the switch is pressed then the LED is shut off.

```
main:
If GPIO.4 = 1 then
HIGH GPIO.0      'Light Red LED
Else
LOW GPIO.0       'Red LED off
ENDIF
```

A GOTO statement completes the loop and sends control back to the main label so the switch can be tested again.

```
GOTO main        'Loop Back to Red LED
```

Software Listing

```
'*****************************************************
'*  Name    : Switch.BAS
'*  Author  : Chuck Hellebuyck
'*  Notice  : Copyright (c) 2009 Electronic Products
'*          : All Rights Reserved
'*  Date    : 8/20/2009
'*  Version : 1.0
'*  Notes   :
'*          : CHIPAXE-8 Pin 1 at C6
'*          : Yellow Jumper - a6 to +rail
'*          : Yellow Jumper - j6 to -rail
'*          : Green Jumper - j7 to j12
'*          : 330 ohm - i12 to i18
'*          : Red LED - Anode j18, Cathode -rail
'*          : Yellow Jumper - j22 to -rail
'*          : Orange Jumper - f22 to e22
'*          : Orange Jumper - b22 to b19
'*          : White Jumper - b8 to b17
'*          : 10k ohm - a17 to +rail
'*          : Switch - d17 to d19
'*****************************************************
```

```
ANSEL = 0  ' Set I/O to digital
CMCON0 = 7 ' Comparator off
input GPIO.4

main:
If GPIO.4 = 1 then
high GPIO.0     'Light Red LED
Else
low GPIO.0      'Red LED off
ENDIF
goto main       'Loop Back to Red LED
```

Next Steps

You could add the LEDs from the traffic light project and change to red, yellow or green on the push of a button. You could also create a speech timer where the green indicates time left is ok, yellow means time is running out and red means times up. The switch can start the process.

You could also just reverse the logic and have the LED light when the switch is pressed. This is a very simple change I'll let you figure out.

Questions

1) If the switch is connected to 5 volts then it is a low side switch.
   A) True
   B) False

2) The IF-THEN-ELSE command line always ends with?
   A) END
   B) ENDIFTHEN
   C) ENDIF
   D) ELSEEND

3) If the TRISIO bit for AN4 is a 0 then the pin is an?
   A) Input
   B) Output
   C) Analog
   D) Not Connected

4) In a high side switch circuit the pull-up resister makes the switch input high when the switch is pressed?
   A) True
   B) False

Chapter 5 - Read a Potentiometer

Many sensors output a signal that is actually a variable voltage or analog voltage. To convert that voltage into a digital value the microcontroller uses an Analog to Digital Converter (ADC). The PIC12F683 has an ADC inside and several of the I/O pins can be connected to the ADC. This just requires some setup in the software. In this project we will create our own adjustable voltage with a potentiometer connected to the GP4 pin. Based on the position of the potentiometer we will light up different LEDs similar to the sound system adjustment display on a stereo system. The completed project is shown in Figure 5-1.

Figure 5-1: Final Potentiometer Project

Hardware

The hardware is similar to other projects with the LEDs wired to three separate ports. The potentiometer is a 10k value that is wired with 5v on one side and ground on the other. The center wiper connection is connected to the GP4 pin of the micro. The connection table describes all the breadboard wiring. The potentiometer has three pins labeled 1, 2 and 3. The center wiper is the number 2 pin. The schematic is shown in Figure 5-2.

Connection Table

```
Micro              - Pin 1 at C6
Yellow Jumper      - a6 to +rail
Yellow Jumper      - j6 to -rail
Green Jumper       - j7 to j12
330 ohm            - i12 to i18
Red LED            - Anode j18, Cathode -rail
Blue Jumper        - g8 to g14
330 ohm            - h14 to h21
Red LED            - Anode j21, Cathode -rail
Purple Jumper      - f9 to f16
330 ohm            - g16 to g24
Red LED            - Anode j24, Cathode -rail
Yellow Jumper      - j22 to -rail
Orange Jumper      - f22 to e22
Yellow Jumper      - b22 to b18
White Jumper       - b8 to b17
Yellow Jumper      - a16 to +
Potentiomter       - 1-e16, 2-c17, 3-e18
```

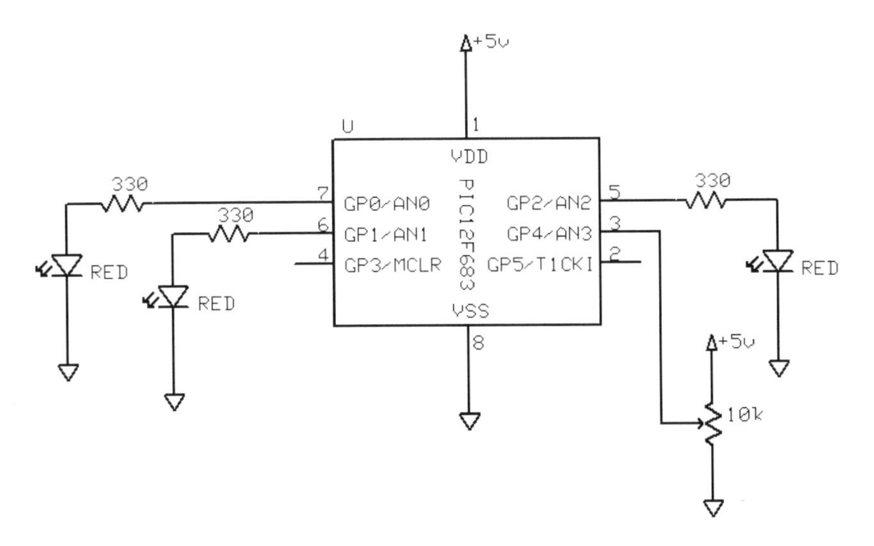

Figure 5-2: Potentiometer Project Schematic

Software

The software starts off differently than the previous projects because we need to setup the analog pins not just the digital pins. The first step is to set the AN4 bit of the ANSEL register to make it analog mode. The rest of the bits are zeros to make the other pins digital.

```
ANSEL = %00001000   ' AN3/GP4 Analog, GP2-GP0 Digital
```

The comparator is still not used so we shut that off with the same CMCON0 setting equal to 7.

```
CMCON0 = 7          ' Comparator off
```

The input/output direction for the pins needs to be setup so we set or clear the bits in the TRISIO register. A one makes the pin and input and a zero makes the pin an output. In this case only the GP4 and GP5 pins are made into an input.

```
TRISIO = %00011000 ' GP4/AN3 input, GP2-GP0 outputs
```

To use the ADC we will use the PICBASIC PRO command ADCIN. This will handle much of the conversion but there are a few things to setup first. All of these items are detailed in the PIC12F683 data sheet but they can be very confusing to a beginner. The values I selected here will work on most projects. The first one we setup is the accuracy. The PIC12F683 can output a 10 bit result (0 to 1024) or an 8 bit result (0 to 256). In this example we will use eight bit mode and we designate that with a DEFINE line shown below.

```
' Define ADCIN parameters
Define ADC_BITS  8   ' Set number of bits in result
```

The ADC uses an internal clock to drive the ADC circuitry. Choosing the proper clock is important when you need a quick conversion but we don't so we use the internal RC option by selecting clock source 3 below with a DEFINE.

```
Define ADC_CLOCK 3   ' Set clock source (3=rc)
```

The ADC charges a capacitor to hold the value while the ADC is converting the voltage. By setting the sample rate we can make sure we allow enough time to make a proper conversion. The DEFINE below handles that.

```
Define ADC_SAMPLEUS 50   ' Set sampling time in uS
```

The ADCIN command will need to store its result in a temporary location so we create a byte variable "adval". If we had used 10 bit mode then we would have to create a word variable because a byte variable can only go up to 255.

```
adval var byte 'Create adval variable to store result
```

Now we enter the main loop at the "main:" label. The first command is the ADCIN command that will measure the voltage on the potentiometer. The GP4 pin is used which is also the AN3 pin. We designate that with the 3 in the command line. The variable we created is also included so the ADCIN command knows where to store the result.

```
main:
ADCIN 3, adval    ' Read channel AN3 to adval
```

The adval variable now has a value from 0 to 255 so we test it using a series of IF-THEN commands to determine which LED to light. If the value is smaller than 90 all three LEDs are lit. If its greater than 90 but less than 180 then two are lit. If the value is greater than 180 but less than 250 then only one is lit. If the value is greater than 250 the LEDs will stay the same as they were and not change.

In each case the LEDs are lit by setting the bit of each LED in the GPIO register. This is a direct way to light the LEDs rather than using several HIGH commands. To use this method we have to make sure the TRISIO register was setup and we did do that earlier in the program.

```
If adval < 90 then   'Test if AN3 is less then 90
GPIO = %00000111     'Light all LEDs
ENDIF
IF adval < 180 then 'Test if AN3 is less than 180
GPIO = %00000011     'Light two LEDS
ENDIF
IF adval < 250 then
GPIO = %00000001     'Light one LED
ENDIF
```

The last command is a GOTO that directs the program back to the main label so the program can test the potentiometer again.

```
goto main            'Loop Back to test potentiometer
```

Software Listing

```
'****************************************************
'*  Name    : Pot.BAS
'*  Author  : Chuck Hellebuyck
'*  Notice  : Copyright (c) 2009 Electronic Products
'*          : All Rights Reserved
'*  Date    : 8/20/2009
'*  Version : 1.0
'*  Notes   :
'*          : Micro      - Pin 1 at C6
'*          : Yellow Jumper - a6 to +rail
'*          : Yellow Jumper - j6 to -rail
'*          : Green Jumper - j7 to j12
'*          : 330 ohm    - i12 to i18
'*          : Red LED    - Anode j18, Cathode -rail
'*          : Blue Jumper - g8 to g14
'*          : 330 ohm    - h14 to h21
'*          : Red LED    - Anode j21, Cathode -rail
'*          : Purple Jumper - f9 to f16
'*          : 330 ohm    - g16 to g24
'*          : Red LED    - Anode j24, Cathode -rail
'*          : Yellow Jumper - j22 to -rail
'*          : Orange Jumper - f22 to e22
'*          : Yellow Jumper - b22 to b18
'*          : White Jumper - b8 to b17
'*          : Yellow Jumper - a16 to +rail
'*          : Potentiomter - 1-e16, 2-c17, 3-e18
'****************************************************

ANSEL = %00001000  ' AN3/GP4 Analog, GP2-GP0 Digital
CMCON0 = 7         ' Comparator off
TRISIO = %00011000 ' GP4/AN3 input, GP2-GP0 outputs

' Define ADCIN parameters
Define ADC_BITS  8  ' Set number of bits in result
Define ADC_CLOCK 3  ' Set clock source (3=rc)
```

```
Define ADC_SAMPLEUS 50  ' Set sampling time in uS

adval var byte 'Create adval variable to store result

main:
ADCIN 3, adval     ' Read channel AN3 to adval

If adval < 90 then  'Test if AN3 is less then 90
GPIO = %00000111    'Light all LEDs
ENDIF
IF adval < 180 then 'Test if AN3 is less than 180
GPIO = %00000011    'Light two LEDS
ENDIF
IF adval < 250 then
GPIO = %00000001    'Light one LED
ENDIF

goto main           'Loop Back to test potentiometer
```

Next Steps

A simple next step is to change the values of the IF-THEN lines to change when to light up the LEDs. You could also change the GPIO arrangement to light the LEDs differently. There is not a lot to change here but if you want to attempt to do an advanced step then try changing to a 10-bit result and give your program more range. The adval variable will need to be a word variable. It should work with just those two changes but then you will probably want to change the IF-THEN values because very little movement of the potentiometer will represent 255.

Questions

1) An ADC converts digital values to analog voltages.
   A) True
   B) False

2) The PIC12F683 can operate in what modes or accuracy?
   A) 6 bit and 8 bit
   B) 8 bit only
   C) 8 bit and 10 bit
   D) 8 bit, 10 bit and 12 bit

3) If the ANSEL bit for a port pin is set to a 1 then that pin is an?
   A) Digital Input
   B) Digital Output
   C) Analog Output
   D) Analog Input

4) A 10 bit resolution requires a variable that is what size?
   A) Byte size
   B) Word size
   C) Byte or Word size
   D) Bit size

Chapter 6 - Sensing Light

In this project we'll use the same Analog to Digital Converter (ADC) to read a CDS cell or light sensing resistor which changes resistance as the light around it changes. Once again an 8-bit resolution result will work perfectly. As the light changes, the ADC value is tested and if it's a high value (high resistance) then its dark out and a single LED lights up. Figure 6 shows the final setup.

Figure 6-1: Final Light Sensor Project

Hardware

The schematic is very similar to the potentiometer schematic except the CDS cell is connected like a fixed resistor. I use a second pull-up resistor to form a resistor divider or voltage divider to convert the changing CDS resistance into a changing voltage. The pull-up resistor could easily be replaced by a potentiometer to give you a sensitivity adjustment. I only drive one LED on or off based on the outside light level similar to a night light. The connection table is below along with the schematic in Figure 6-2.

Connection Table

```
Micro         Pin 1 at C6
Yellow Jumper - a6 to +rail
Yellow Jumper - j6 to -rail
Green Jumper - j7 to j12
330 ohm - i12 to i18
Read LED - Anode j18, Cathode -rail
Yellow Jumper - j22 to -rail
Orange Jumper - f22 to e22
Yellow Jumper - b22 to b18
White Jumper - b8 to b17
1k ohm - a17 to +rail
CDS Cell - d17 to d18
```

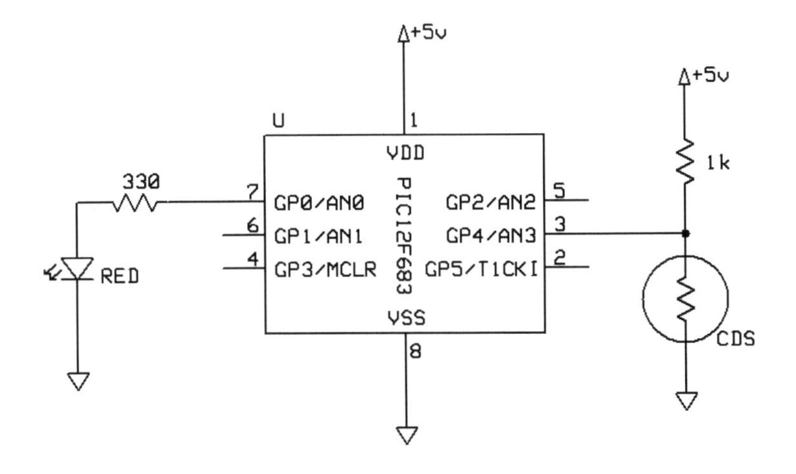

Figure 6-2: Light Sensor Project Schematic

Software

The start of the software is the same as the potentiometer setup since we will be using the GP4 pin as an analog input. Using the binary bit designation in PICBASIC PRO allows me to easily set the AN4 bit of the ANSEL register making it analog while the rest of the I/O pins are zero or set to digital mode.

```
ANSEL = %00001000  ' AN3/GP4 Analog, GP2-GP0 Digital
```

The comparator is shut down once again.

```
CMCON0 = 7 ' Comparator off
```

The state of the GP4 has to be set to input mode using the TRISIO register which can also set the rest of the I/O pins are set to outputs.

```
TRISIO = %00011000 ' GP4 input, GP2 thru GP0 outputs
```

The ADCIN command requires some setup parameters to be established such as the ADC resolution, ADC clock source and sampling time. These are easily done with DEFINE statements.

```
' Define ADCIN parameters
Define ADC_BITS  8 ' Set number of bits in result
Define ADC_CLOCK 3      ' Set clock source (3=rc)
Define ADC_SAMPLEUS 50  ' Set sampling time in uS
```

A variable is established to store the ADCIN result.

```
adval var byte 'Create adval variable to store result
```

The main label establishes the main loop followed by the ADCIN command line where the CDS is read.

```
main:
ADCIN 3, adval      ' Read channel AN3 to adval
```

After the value of the CDS cell is stored in the adval variable it is compared to the value 150. If it is less than 150 then the LED is off. If the value is greater than 150 then it's dark and the LED is lit using the HIGH command.

```
If adval > 150 then      'Light LED if in the dark
High GPIO.0              'Light all LEDs
ELSE
LOW GPIO.0
ENDIF
```

Another GOTO statement completes the main loop.

```
goto main    'Loop Back to test potentiometer
```

Software Listing

```
'****************************************************
'*  Name    : CDS.BAS
'*  Author  : Chuck Hellebuyck
'*  Notice  : Copyright (c) 2009 Electronic Products
'*          : All Rights Reserved
'*  Date    : 8/20/2009
'*  Version : 1.0
'*  Notes   :
'*          : CHIPAXE-8 Pin 1 at C6
'*          : Yellow Jumper - a6 to +rail
'*          : Yellow Jumper - j6 to -rail
'*          : Green Jumper - j7 to j12
'*          : 330 ohm - i12 to i18
'*          : Read LED - Anode j18, Cathode -rail
'*          : Yellow Jumper - j22 to -rail
'*          : Orange Jumper - f22 to e22
'*          : Yellow Jumper - b22 to b18
'*          : White Jumper - b8 to b17
'*          : 1k ohm - a17 to +rail
'*          : CDS Cell - d17 to d18
'****************************************************

ANSEL = %00001000  ' AN3/GP4 Analog, GP2-GP0 Digital
CMCON0 = 7         ' Comparator off
TRISIO = %00011000 ' GP4 input, GP2 thru GP0 outputs

   ' Define ADCIN parameters
Define ADC_BITS  8     ' Set number of bits in result
Define ADC_CLOCK  3      ' Set clock source (3=rc)
Define ADC_SAMPLEUS 50   ' Set sampling time in uS

adval var byte 'Create adval variable to store result

main:
```

```
ADCIN 3, adval          ' Read channel AN3 to adval

If adval > 150 then     'Light LED if in the dark
High GPIO.0             'Light all LEDs
ELSE
LOW GPIO.0
ENDIF

goto main             'Loop Back to test light sensor
```

Next Steps

Changing the threshold value from 150 to something higher or lower will determine how dark it has to be to light the LED. The limit is 0 to 255 since we used an 8-bit result. Just like the switch project other sensors can replace the light sensor. A thermistor could replace the light sensor to measure temperature. A potentiometer could be used to create a manual interface. If you remove the pull-up resistor then a Sharp GP2D12 Object Detection Sensor that produces a variable output voltage can be read by the analog pin directly for more accurate robotic obstacle detection.

Questions

1) A CDS Senses temperature.
   A) True
   B) False

2) When the light is bright on the CDS cell the resistance is?
   A) High
   B) Low
   C) Zero
   D) Infinite

3) The pull-up resistor connected to the CDS forms a?
   A) High Side Switch
   B) Low Side Switch
   C) Voltage Divider
   D) Potentiometer

4) A 8 bit resolution requires a variable that is what size?
   A) Byte size
   B) Word size
   C) Byte or Word size
   D) Bit size

Chapter 7 – Creating Sound

There are many electronic gadgets that beep when you press a button. This is known as audible feedback. We will recreate that sound here with a piezo speaker. The PIC12F683 will sense a momentary switch press with a digital input. When the switch is pressed the software will generate a square wave at a fixed frequency in the audible range. The piezo speaker will generate the tone for a short time and then wait for the switch to be pressed again.

Figure 7-1: Final Creating Sound Project

Hardware

The hardware replaces the light sensor in the previous project with a normally open momentary switch connected to the GP4 pin and using a 10k pullup resistor. In this example we'll make GP4 a digital port not an analog port as we did in the light sensor project. The GP2 pin is configured as a digital output that creates a square wave pulse using the software SOUND command. The square wave needs to be converted into a rounded signal to works best with the piezo speaker so we place a 10 uf capacitor in series between the GP2 pin and the Piezo positive lead. The piezo speaker is then grounded. The connection table describes how to connect all this on the breadboard. The Piezo speaker is from Jameco.com under part number DBX05-PN.

Connection Table

```
CHIPAXE-8 Pin 1 at C6
Yellow Jumper - a6 to +rail
Yellow Jumper - j6 to -rail
Orange Jumper - j9 to j12
10uf 35v - Positive-h12, Negative-h13
Piezo - Positive-g17, Negative-i19
Yellow Jumper - f13 to f17
Yellow Jumper - j19 to -rail
Yellow Jumper - j22 to -rail
Orange Jumper - f22 to e22
Orange Jumper - b22 to b19
White Jumper - b8 to b17
10k ohm - a17 to +rail
Switch - d17 to d19
```

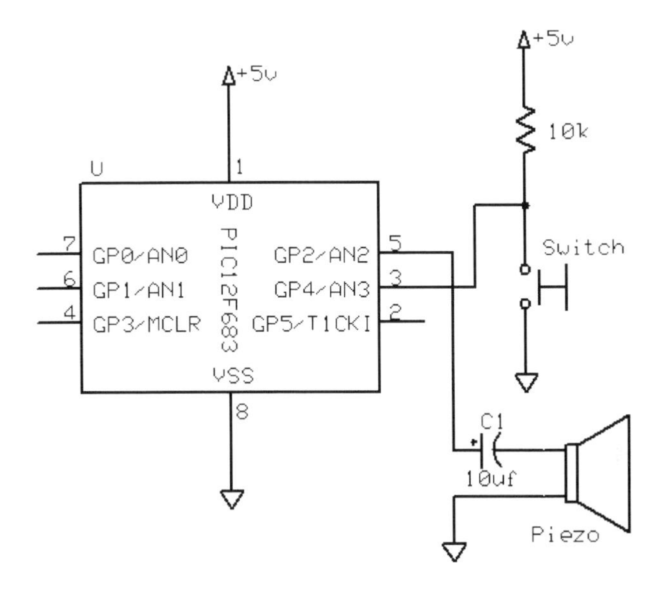

Figure 7-2: Light Sensor Project Schematic

Software

The software takes full advantage of the SOUND command in PICBASIC PRO. This command creates the square wave that drives the piezo speaker. To get started we need to make the pins digital by clearing the ANSEL register.

```
ANSEL = 0        ' Set I/O to digital
```

The comparator is turned off as usual by setting the CMCON0 register to seven.

```
CMCON0 = 7       ' Comparator off
```

The GP4 pin will sense the switch so we need to make sure it is an input by using the input command followed by pin name. I use the optional port.pinname format.

```
input GPIO.4   ' Make GP4 and input
```

The main loop of code starts with the familiar "main:" label.

```
main:
```

The If-Then command is used to test the GP4 pin for a low state which indicates the switch has been pressed.

```
If GPIO.4 = 0 then 'If switch pressed create sound
```

If the switch is pressed then the SOUND command creates a square wave for a period of time. The first value has be a value between 1 and 127. The value 1 produces a signal of about 78Hz and 127 produces 10,000 Hz so the 100 value is somewhere between those. The second value is the duration which can be between 1 and 255. Each value represents about a 12 millisecond duration. In this case the sound lasts about 1.2 seconds.

```
sound GPIO.2,[100,100] 'Sound speaker for 1/10 sec
```

The If-Then command ends with the ENDIF command.

```
ENDIF
```

The GOTO command finishes the main loop of code by jumping the program control back to the main: label.

goto main 'Loop Back to test the switch

Software Listing

```
'**************************************************
'*  Name    : Sound.BAS
'*  Author  : Chuck Hellebuyck
```

```
'*  Notice  : Copyright (c) 2009 Electronic Products
'*          : All Rights Reserved
'*  Date    : 8/20/2009
'*  Version : 1.0
'*  Notes   :
'*          : CHIPAXE-8 Pin 1 at C6
'*          : Yellow Jumper - a6 to +rail
'*          : Yellow Jumper - j6 to -rail
'*          : Orange Jumper - j9 to j12
'*          : 10uf 35v - Positive-h12, Negative-h13
'*          : Piezo - Positive-g17, Negative-i19
'*          : Yellow Jumper - f13 to f17
'*          : Yellow Jumper - j19 to -rail
'*          : Yellow Jumper - j22 to -rail
'*          : Orange Jumper - f22 to e22
'*          : Orange Jumper - b22 to b19
'*          : White Jumper - b8 to b17
'*          : 10k ohm - a17 to +rail
'*          : Switch - d17 to d19
'*********************************************

ANSEL = 0        ' Set I/O to digital
CMCON0 = 7       ' Comparator off
input GPIO.4     ' Make GP4 and input

main:
If GPIO.4 = 0 then 'If switch pressed create sound
sound GPIO.2,[100,100] 'Sound speaker for 1/10 sec
ENDIF

goto main        'Loop Back to test the switch
```

Next Steps

There are many options to this simple program. You can make the piezo sound last longer or shorter by changing the values in the SOUND command. You can make the sound occur constantly unless the switch is pressed. This would be handy if you needed to keep an eye on something. Have it sit on top of the switch so if someone lifts up the item then the alarm will go off. You could also add another switch so you have one turn on the sound and the other turn it off. This could be a fun project to modify.

Questions

1) An If-Then command can be used to test a switch?
   A) True
   B) False

2) A SOUND command produces a sine wave signal?
   A) True
   B) False

3) The pull-up resistor on a low side switch makes the idle state?
   A) High
   B) Low
   C) High Impedance
   D) Normally Open

4) Sound (100,200) will produce a sound for?
   A) 100 milliseconds
   B) 20 milliseconds
   C) 200 milliseconds
   D) 10 milliseconds

Chapter 8 – Sensing Vibration

The next project can be very useful for measuring impact or vibration of an object. The project uses a flexible piezo type vibration sensor that produces a voltage when vibration is sensed. By reading that voltage with the analog to digital converter inside the PIC12F683 the software can indicate vibration is present by driving a piezo speaker to beep. If the circuitry were attached to an item this could be used as an alarm system to sense when an object is moved. It's sensitive enough to also sense foot steps. I've always wondered how much vibration a package would see when shipped through the Post Office. Maybe this will form the basis for that type of project in the future.

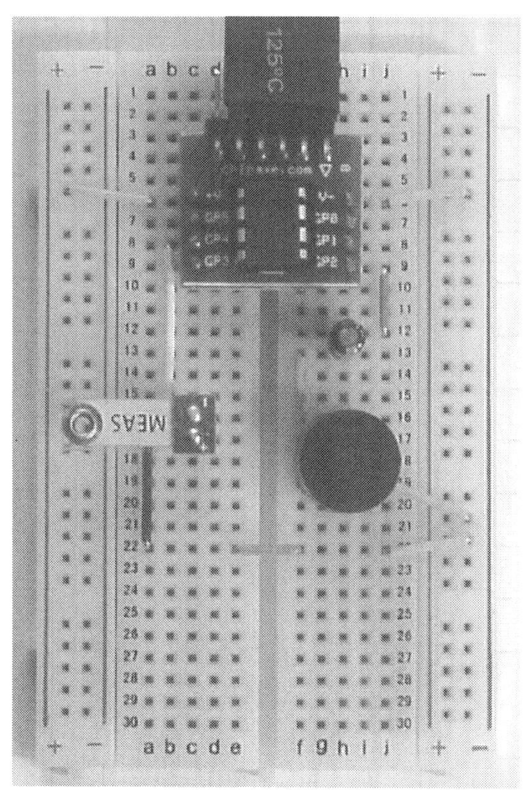

Figure 8-1: Final Sensing Vibration Project

Hardware

The hardware consists of a CHIPAXE module positioned in the center of the breadboard with jumpers to the power rails so the programmer can power the whole circuit. The project uses a small vibration sensor from Measurement Specialties part number 1005939-1 (see Appendix A for source) that produces a voltage when it's vibrating. The voltage output is fed directly into the GP4 pin which is setup in the software as an analog input the same way we did in the light sensor project.

The GP2 pin is configured as a digital output that creates a square wave pulse using the software SOUND command to drive a piezo speaker. As before, the square wave needs to be converted into a rounded signal to work best with the piezo speaker so we place a 10 uf capacitor in series between the GP2 pin and the Piezo positive lead. The piezo speaker is then grounded. The piezo used is a DBX05-PN speaker from Jameco electronics. The connection table describes how to connect all this on the breadboard and Figure 8-2 shows the schematic.

One thing to note in the picture is the vibration sensor looks like it's in backwards with the positive pin connected to ground. This is not the case. The jumpers actually cross each other underneath the sensor to make the proper connection of negative to ground. Sometimes pictures don't tell the whole story but the connection table below does.

Connection Table

```
CHIPAXE-8 Pin 1 at C6
Yellow Jumper - a6 to +rail
Yellow Jumper - j6 to -rail
Orange Jumper - j9 to j12
10uf 35v - Positive-h12, Negative-h13
Speaker - Positive-g17, Negative-i19
Yellow Jumper - f13 to f17
Yellow Jumper - j19 to -rail
```

```
Yellow Jumper - j22 to -rail
Orange Jumper - f22 to e22
Blue Jumper - a22 to a16
White Jumper - b8 to b17
Vibration Sensor - Pos-d17, Neg-d16
```

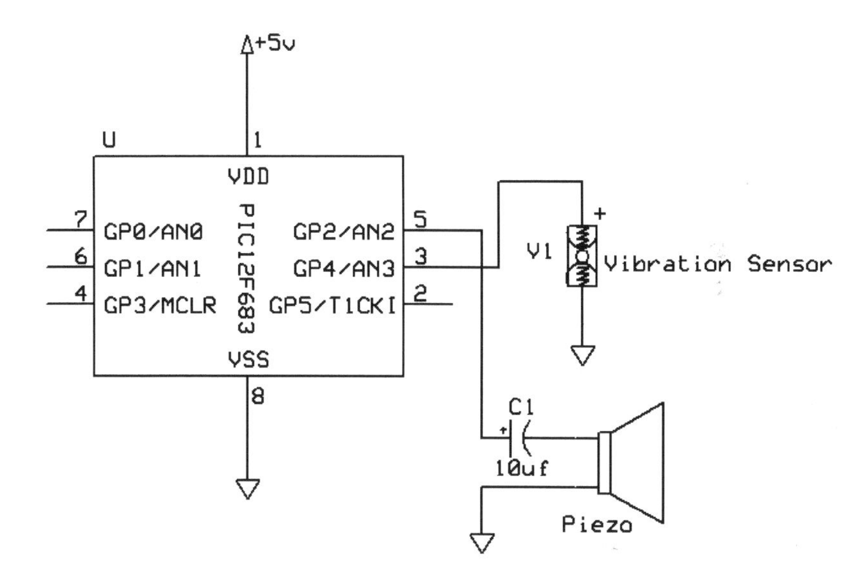

Figure 8-2: Vibration Sensor Project Schematic

Software

The software is very similar to previous projects as it uses many of the same commands. We use the SOUND command and the ADCIN command once again. To get started we need to make the speaker pin digital except for the GP4 pin which will read the vibration sensor. We can set each bit by using the percent symbol in front of the 1's and 0's so the compiler knows we want to use a binary number to set each bit. In this case all unused pins are also setup as digital pins.

```
ANSEL = %00001000 'I/O digital except AN3/GP4 Analog
```

The comparator is turned off as usual by setting the CMCON0 register to seven.

```
CMCON0 = 7       ' Comparator off
```

The GP4 pin will sense the vibration sensor so we need to make sure it is an input by using the binary method to set all the pins state in the TRISIO register. A one makes a digital pin an input and a zero makes a digital pin an output. Even though we made GP4 an analog pin we still have to make it an analog input. This seems redundant but is required inside the microcontroller. An analog pin can be made into an output but it won't work properly.

```
TRISIO = %00011000 ' GP4 input, GP2 thru GP0 outputs
```

The analog to digital settings for the ADCIN command are established with a set of DEFINE statements. These are the default settings so this step can be skipped but I like to define them in the code so I have the settings in writing. These do take up some of the 31 command limit so if you run out of space with a program based on these lines then you can comment them out, save command lines and still know what the default settings are.

```
' Define ADCIN parameters
Define ADC_BITS 8   ' Set number of bits in result
Define ADC_CLOCK 3 ' Set clock source (3=rc)
Define ADC_SAMPLEUS 50   ' Set sampling time in uS
```

The ADCIN command needs to store the vibration value in a variable so we have to create a variable for that. That is done in the line below.

```
adval var byte  ' Create adval to store result
```

The main loop of code starts with the familiar main: label.

```
main:
```

The ADCIN command is the first step and reads the GP4 pin which is also the AN3 pin or analog pin 3. The result is stored in the adval variable.

```
ADCIN 3, adval  ' Read channel AN3 to adval
```

The next step is to test the value to see if vibration is present. This is where you set the sensitivity for the vibration sensor. The voltage of the vibration sensor will result in an ADC read of 0 to 255 since we are using eight bit mode (set back in the Define command lines). If the voltage results in a value larger than 150 then a beep will be produced in the speaker. If you want to make it more sensitive, then lower this value. If you want to make less sensitive than raise the value.

```
If adval > 150 then   'Light LED if in the dark
```

If the vibration is large enough then the SOUND command creates a square wave for a period of time. The first value has be a value between 1 and 127. The value 1 produces a signal of about 78Hz and 127 produces 10,000 Hz so the 100 value is somewhere between those. The second value is the duration which can be between 1 and 255. Each value represents about a 12 millisecond duration. In this case the sound lasts about 1.2 seconds.

```
sound GPIO.2,[100,100] 'Sound speaker for 1/10 sec
```

The If-Then command ends with the ENDIF command.

```
ENDIF
```

The GOTO command finishes the main loop of code by jumping the program control back to the main: label.

```
goto main      'Loop Back to test the vibration sensor
```

Software Listing

```
'****************************************************
'*  Name    : Vibration.
'*  Author  : Chuck Hellebuyck
'*  Notice  : Copyright (c) 2009 Electronic Products
'*          : All Rights Reserved
'*  Date    : 8/20/2009
'*  Version : 1.0
'*  Notes   :
'*          : CHIPAXE-8 Pin 1 at C6
'*          : Yellow Jumper - a6 to +rail
'*          : Yellow Jumper - j6 to -rail
'*          : Orange Jumper - j9 to j12
'*          : 10uf 35v - Positive-h12, Negative-h13
'*          : Speaker - Positive-g17, Negative-i19
'*          : Yellow Jumper - f13 to f17
'*          : Yellow Jumper - j19 to -rail
'*          : Yellow Jumper - j22 to -rail
'*          : Orange Jumper - f22 to e22
'*          : Blue Jumper - a22 to a16
'*          : White Jumper - b8 to b17
'*          : Vibration Sensor - Pos-d17, Neg-d16
'****************************************************

ANSEL = %00001000  'I/O digital except AN3/GP4 Analog
CMCON0 = 7 ' Comparator off
TRISIO = %00011000 ' GP4 input, GP2 thru GP0 outputs

' Define ADCIN parameters
Define ADC_BITS 8  ' Set number of bits in result
Define ADC_CLOCK 3 ' Set clock source (3=rc)
Define ADC_SAMPLEUS 50  ' Set sampling time in uS

adval var byte  ' Create adval to store result

main:
ADCIN 3, adval  ' Read channel AN3 to adval

If adval > 150 then   'Light LED if in the dark
sound GPIO.2,[100,100] 'Create sound for 1/10 sec
ENDIF

goto main       'Loop Back to test vibration sensor
```

Next Steps

This type of vibration sensing setup could be used in many applications where you need to test for movement. The sensor operates best in an up and down direction so you might want to add a second sensor mounted at a 90 degree angle to the first one for another angle of sensing. Measurement Specialties makes the same sensor with a connector designed to mount at 90 degrees. That would be a great next step.

As I mentioned at the beginning of this chapter, this setup could also be used to measure vibration on any item including being shipped in a box to see how well the Post Office handles the package. To do that you would need to store the value in memory and that requires an EEPROM or Electrically Erasable Programmable Read Only Memory which is another topic for another book. Besides I don't think the post office will allow a package with electronics running inside to be shipped so I may never get a chance to test that idea.

Questions

1) A % symbol in front of 1' and 0's represents what number system?

A) Octal
B) Binary
C) Hexadecimal
D) Decimal

2) A piezo element in any sensor produces?

A) High Signal
B) Low Signal
C) Variable Voltage
D) Variable Sound

3) What is largest range an eight bit analog to digital result can have?

A) 0 – 128
B) 255- 1024
C) 0-1024
E) 0 - 255

4) The capacitor in series with the piezo speaker converts an analog signal into a digital signal?

A) True
B) False

Conclusion

The number of projects you can complete with the PIC12F683 and PICBASIC PRO sample version is really endless. I only touched the surface of what you can do with this little eight pin microcontroller which leads me to consider a second book with more projects. The projects in this book can easily be expanded to do more complex projects but you will probably need the full version of the PICBASIC PRO compiler. Clearly though this book was written for the person just getting started with programming. The CHIPAXE programming system makes this easy and low cost but all these projects could easily be adapted to the PICkit 2 Starter Kit from Microchip as well.

Learning how to use digital inputs and outputs along with analog to digital converter inputs is essential for the beginner. These functions will probably be part of every project you tackle beyond this book. I also didn't want to put too much in this book so the reader didn't feel they have to learn too many topics to get started. Sometimes the feeling of accomplishment by completing one book is a great motivator to move on to tougher subjects in another. This is why this book is purposefully short but to the point. It also allows me to offer it at a lower price so more people will be willing to purchase the book and get started. If you team this book up with the CHIPAXE programmer set then you can be programming for well under $100. In fact I plan to write more books for the PIC12F683 and the CHIPAXE to create a series of books to learn from. Keep an eye out for them in the future.

Speaking of books, if you decide to move on to the C language then I hope you get a chance to check out my books: "Beginner's Guide to Embedded C Programming" Volumes 1 and 2. These books shows how to use many of the PIC16F690 peripherals and also how to drive LCDs, LED displays and other common functions using the HI-TECH C compiler. These are available from my website www.elproducts.com.

I hope you learned a lot from this book and had fun programming. If you have any questions or comments, please don't hesitate to send them to me at chuck@elproducts.com.

Appendix A –Parts List for Projects

1 – CHIPAXE 8-pin Breadboard Starter Kit (CHIPAXE.com #CHIP002)
1 – Normally Open Momentary Switch (Jameco.com #199726)
1 – Vibration Sensor 1005939-1 (Digikey.com #MSP6914-ND)
1 – Piezo Speaker DBX05-PN (Jameco.com #138740)
1 – CDS Photocell (Jameco #202366)
1 – 10k Trim Potentiometer (Jameco #43001)
1 – 10uf 16v Electrolytic Capacitor (Jameco #198839)
1 – Red Diffused LED T-1 ¾ (Jameco #333973 min qty 10 pcs)
1 – Yellow Diffused LED T-1 ¾ (Jameco #333622 min qty 10 pcs)
1 – Green Diffused LED T-1 ¾ (Jameco #253833 min qty 10 pcs)
3 – 330 Ohm ¼ w Resistor (Jameco #690742 min qty 100 pcs)
1 – 1k ¼ w Resistor (Jameco #690865 min qty 100 pcs)
1 – 10k ¼ w Resistor (Jameco #691104 min qty 100 pcs)

A complete kit of parts is also available from
www.CHIPAXE.com.

All the software used in this book can be downloaded from:
www.elproducts.com/eightpinbook_vol1.htm

Appendix B – Answers to Questions

Chapter 2

1) B
2) C
3) B
4) D

Chapter 3

1) D
2) B
3) C
4) A

Chapter 4

1) B
2) C
3) B
4) A

Chapter 5

1) B
2) C
3) D
4) B

Chapter 6

1) B
2) B
3) C
4) A

Chapter 7

1) A
2) B
3) A
4) C

Chapter 8

1) B
2) C
3) D
4) B

Index

Made in the USA
Charleston, SC
22 March 2016